Smart Cities and the UN SDGs

Edited by

Anna Visvizi
SGH Warsaw School of Economics, Warsaw, Poland
Effat University, Jeddah, Saudi Arabia

Raquel Pérez del Hoyo
University of Alicante, Alicante, Spain

ELSEVIER

Elsevier
Radarweg 29, PO Box 211, 1000 AE Amsterdam, Netherlands
The Boulevard, Langford Lane, Kidlington, Oxford OX5 1GB, United Kingdom
50 Hampshire Street, 5th Floor, Cambridge, MA 02139, United States

Library of Congress Cataloging-in-Publication Data
A catalog record for this book is available from the Library of Congress

British Library Cataloguing-in-Publication Data
A catalogue record for this book is available from the British Library

ISBN: 978-0-323-85151-0

For information on all Elsevier publications
visit our website at https://www.elsevier.com/books-and-journals

Publisher: Joe Hayton
Acquisitions Editor: Brian Romer
Editorial Project Manager: Alice Grant
Production Project Manager: Swapna Srinivasan
Cover Designer: Christian Bilbow

Typeset by SPi Global, India

Smart Cities and the UN SDGs

Contents

14. Smart city as the city of knowledge

Robert Laurini, Giuseppe Las Casas, and Beniamino Murgante

15. Foresight: A visionary step for becoming a smart city

Andrei Ștefan Neștian and Ana Iolanda Vodă

16. Transforming the city from within: SDGs and their transformative dynamics in the smart city (conclusions)

Anna Visvizi and Raquel Pérez del Hoyo

Contributors

Numbers in parenthesis indicate the pages on which the authors' contributions begin.

Shahira Assem Abdel-Razek (61), Delta University for Science and Technology, Department of Architectural Engineering, Dakahlia, Egypt

Roberto O. Andrade (181), School of Systems Engineering, National Polytechnic School, Quito, Ecuador

Francesco Bifulco (29), University of Naples Federico II, Naples, Italy

Anna D'Auria (29), University of Naples Federico II, Naples, Italy

Raquel Pérez del Hoyo (1, 15, 107, 253), University of Alicante, Alicante, Spain

Clara García-Mayor (107), University of Alicante, Alicante, Spain

Vasiliki Geropanta (143), Technical University of Crete, Chania, Greece

Pablo de Gracia-Soriano (197), Complutense University of Madrid, Madrid; University of Alicante, Alicante, Spain

Mara Grimaldi (71), Department of Management & Innovation Systems, University of Salerno, Fisciano, Italy

Diana Jareño-Ruiz (197), University of Alicante, Alicante, Spain

Antonio Jiménez-Delgado (197), University of Alicante, Alicante, Spain

María Jiménez-Delgado (197), University of Alicante, Alicante, Spain

Anna Karagianni (143), Technical University of Crete, Chania, Greece

Giuseppe Las Casas (211), LISUT, School of Engineering, University of Basilicata, Potenza, Italy

Robert Laurini (211), Knowledge Systems Institute, Skokie, IL, United States; LIRIS, University of Lyon, Lyon, France

Francesca Loia (71), Department of Economics, Management and Institutions University of Naples 'Federico II', Naples, Italy

Piotr Maleszyk (163), Maria Curie-Skłodowska University, Lublin, Poland

Pablo Martí (107), University of Alicante, Alicante, Spain

Sofia Mavroudi (143), Technical University of Crete, Chania, Greece

Christine Meschede (91), Department of Information Science, Heinrich Heine University Düsseldorf, Düsseldorf, Germany

Higinio Mora (15), University of Alicante, Alicante, Spain

Beniamino Murgante (211), LISUT, School of Engineering, University of Basilicata, Potenza, Italy

Andrei Ştefan Neştian (233), Faculty of Economics and Business Administration, Alexandru Ioan Cuza University, Iaşi, Romania

Almudena Nolasco-Cirugeda (107), University of Alicante, Alicante, Spain

Iván Ortiz-Garcés (181), Faculty of Engineering and Applied Sciences, University of the Americas, Quito, Ecuador

Chrysaida-Aliki Papadopoulou (45), School of Rural and Surveying Engineering, National Technical University of Athens, Athens, Greece

Panagiotis Parthenios (143), Technical University of Crete, Chania, Greece

Francesco Polese (71), Department of Management & Innovation Systems, University of Salerno, Fisciano, Italy

Cláudia Toriz Ramos (125), University Fernando Pessoa & CEPESE, Porto, Portugal

Leticia Serrano-Estrada (107), University of Alicante, Alicante, Spain

Tobias Siebenlist (91), Ronin Institute for Independent Scholarship, Montclair, NJ, United States

Luis Tello-Oquendo (181), College of Engineering, Universidad Nacional de Chimborazo, Riobamba, Ecuador

Marco Tregua (29), University of Naples Federico II, Naples, Italy

Orlando Troisi (71), Department of Management & Innovation Systems, University of Salerno, Fisciano, Italy

Anna Visvizi (1, 15, 253), SGH Warsaw School of Economics, Warsaw, Poland; Effat University, Jeddah, Saudi Arabia; Deree College, The American College of Greece, Athens, Greece

Ana Iolanda Vodă (233), Institute for Interdisciplinary Research, Social Science and Humanities Research Department, Alexandru Ioan Cuza University, Iaşi, Romania

Sang Guun Yoo (181), School of Systems Engineering; Smart Lab, National Polytechnic School, Quito, Ecuador

Acknowledgments

This volume is marked by the Covid-19 pandemic. The initial proposal was written when several countries in the European Union (EU) where already in lockdown. The final version of the book was submitted when the second wave of the pandemic placed several countries in the EU under lockdown again. All of us are aware of the anxiety, uncertainty, and exhaustion that this situation infuses in our professional and personal lives. Therefore, we are particularly grateful to all and to everyone who were part of this project from day one. We would like to thank all contributing authors who addressed our endless queries and suggestions regarding their chapters. We would also like to thank the anonymous reviewers of this book, as their detailed and, yet very friendly, feedback allowed us to reconsider certain assumptions and arguments. We greatly appreciate it. We would also like to thank the publisher, and especially the Executive Editor, Brian Romer, whose guidance and support were quintessential for the completion of this project. Finally, we would like to thank the entire Elsevier's editorial team. We recognize that all parties that contributed to this book worked under an enormous stress. We would like to thank you all for your commitment and trust.

The Editors
Anna Visvizi
Raquel Pérez del Hoyo

Chapter 1

Sustainable development goals (SDGs) in the smart city: A tool or an approach? (An introduction)

Anna Visvizi[a,b] and Raquel Pérez del Hoyo[c]
[a]*SGH Warsaw School of Economics, Warsaw, Poland,* [b]*Effat University, Jeddah, Saudi Arabia,* [c]*University of Alicante, Alicante, Spain*

1. Introduction

This edited volume looks at smart cities and the smart city debate through the lens of the sustainable development goals (SDGs) and the broader discussion on sustainability as it unfolds in research and popular discourse worldwide. By engaging in this discussion, the authors contributing to this volume explore how the encounter of sophisticated technology and progressing urbanization may yield positive results in city/urban space. The measure of success is defined by the, entailed in the SDG 11 four-pronged goal of inclusion, safety, resilience, and sustainability. The SDGs have defined the frames of the policy-making process and business strategy development around the world. Smart cities represent in this context a specific case and this book explores it in depth. A wealth of literature on smart cities exist (Anthopoulos, 2019; Cardullo et al., 2019; Min et al., 2020; Vacca, 2020; Visvizi & Lytras, 2019a; Yamagata & Yang, 2020). Each of these contributions, adopting a specific disciplinary perspective and promoting a specific worldview, adds a piece of knowledge and understanding of the smart city and its role in the ceaseless reconstitution of our societies (van Soest et al., 2019; Wu, Chen, Wu, & Lytras, 2018). This volume seeks to add to this debate.

The impact of information and communication technology (ICT) on our societies is pervasive. More frequently than not, ICT and ICT-enhanced services and solutions are applied in the public and private domains even before the average citizen recognizes their existence. A dense web of distributed and interconnected ICT-enhanced services and applications emerges and ties the fabric of the city in a manner unknown before. A new virtual dimension is added

Smart Cities and the UN SDGs. https://doi.org/10.1016/B978-0-323-85151-0.00001-4

to the city space enabling day-to-day public services delivery, business activity, as well as representation and participation in issues pertinent to the city (Marvin, Luque-Ayala, & McFarlane, 2015; OECD, 2020b; Visvizi & Lytras, 2019b). Simultaneously, the definition of the open, tangible public space, traditionally defining the city alters as we speak. Substantial challenges and opportunities emerge for the field of urban design. As the inroads of ICT-enhanced applications and services continue, it is necessary to address the persistent duality of smart city and sustainable city. This book seeks to do just that. The collection of papers included in this volume makes a case that a city can only be truly smart if the promise and the potential that the ICT brings are employed to promote inclusion, safety, resilience, and sustainability. Several dimensions of these concepts exist (Clarinval et al., 2020; Gámez-García et al., 2019; Kitchin et al., 2019; Moreno-Ibarra & Torres-Ruiz, 2019; Shmelev, 2019; Zait, 2017) and each of these should be taken into consideration when examining specific aspects of the smart city and its development and functioning. The objective of this chapter is to shed light on the concepts of the smart city and the SDGs to offer a guided insight into the book's rationale and content. Conclusions follow.

2. Connecting the smart city and the SDGs

Smart cities and SDGs are two potent policy-making paradigms that shape the thrust of academic and popular debates. Launched in 2015, the SDGs were thought to build on the heritage of the Millennium Development Goals (MDGs). While the MDGs had been discussed in terms of a reasonable success story (Assefa et al., 2017; Fukuda-Parr et al., 2013; Ruhil, 2015), the implementation of the SDGs seems to be sluggish, thus giving ground to voices of criticism (Gusmão Caiado et al., 2018; Ramirez-Rubio et al., 2019). The SDGs represent a major multilateral effort to shift the stakeholders' attention toward more sustainable and resilient pathways of growth and development. Considerable body of literature exists in which the intricacies of the SDGs are discussed (Bali Swain & Yang-Wallentin, 2020; Christ & Burritt, 2019; Moyer & Hedden, 2020; Vandemoortele, 2018). The important, and yet frequently unrecognized, feature of the SDGs is that the very definition of respective goals forming the SDG agenda, one, inserts the developed countries in the scope of the SDGs, and, two, makes it easier for diverse stakeholders to embark on the implementation of specific goals in line with their organization's mission and vision. This, seemingly tiny, yet audit-wise important, issue allows companies to devise budget lines geared toward targeted implementation of specific SDGs. This flexibility ingrained in the SDGs partly explains why it is difficult to measure progress toward the SDGs' implementation (cf. Ali et al., 2018; Leal Filho, 2020; Miola & Schiltz, 2019). In the context of the debate on the SDG 11 and cities, the key problem that this volume identifies is the following. In essence, the imperative of making cities inclusive, safe, resilient, and sustainable implicitly imposes the responsibility for the goal's attainment on the local authorities, say municipalities. It is not always, however, the case that the local authorities have the legal powers, resources, or know-how to implement respective solutions. In this regard, progress toward the four-pronged goal of inclusion, safety, resilience, and sustainability is doomed to be uneven and asynchronous, both at international and local levels.

Progressing urbanization, the increasing density of population in the limited city space, however, also the condensation of talent in the geographically constrained city space, require that the question of power assigned to the city authorities is reconsidered. Certainly, for quite some time now, substantial effort was invested in this issue, mostly through policies and regulatory frameworks aimed at decentralization of power (OECD, 2019), promotion of local (self) governance (UNDP, 2010), and multilevel governance initiatives (OECD, 2017). Still more effort is needed on the part of national governments and city authorities to devise strategies and action plans aimed at the implementation of the SDG11 in the city space. The case of Bonn (Germany) is a welcome highlight in this respect. With the support of the Federal Ministry for Economic Cooperation and Development, the city authorities implement a strategy aimed at providing affordable housing, expanding and maintaining green spaces, shifting to clean forms of transport and energy, and providing employment opportunities for all, especially for low-skilled workers. Ways of managing trade-offs between climate, sustainable mobility, and affordable housing goals form a part of the strategy too (OECD, 2020a). Several other examples would be possible (cf. Carroll, 2020; Gobin-Rahimbux et al., 2020).

This volume stems from the recognition that the smart city, seen as the key concept in the policy-making and urban planning agendas, faces a variety of challenges. These can be identified at several levels, including infrastructure; citizens' ability and willingness to use smart city services, including the question of accessibility; citizen participation and representation; the increasingly salient issue of data privacy; data sovereignty; and many more. These issues have been highlighted in the academic debate on smart cities (Calzada & Almirall, 2020; Lytras, Visvizi, Chopdar, Sarirete, & Alhalabi, 2021; Lytras & Visvizi, 2018; Mora & Deakin, 2019). This volume expands the insights contained therein, frequently by means of adding the empirical and comparative perspectives.

In the context defined by an increasing pace of urbanization, especially in so-called Global South, by the increasing recognition and the speed of application of ICT-enhanced services and applications in the city, this volume sheds light on the promise and constraints related to the attainment of the four-pronged goal of inclusion, safety, resilience, and sustainability in the smart city. The SDGs have been promoted as a comprehensive approach to tackle the key issues and challenges that inhibit sustainable and growth and development today. Against the backdrop of criticisms voiced vis-à-vis the SDGs, this volume suggests that the SDGs be viewed as a broad and a long-term strategy. Attempts to quantify the progress in the SDGs are useful but misleading. The value of the SDGs is hidden in-between the numbers. It is best represented by the change induced in citizens' and decision-makers' attitudes to planning and policy-making. The smart city is but one of the domains where this change takes place. This book sheds light on it.

3. Review of the chapters' content

The book's content is divided into four parts. Part 1 offers a detailed insight into the definitional and conceptual issues related to the implementation of the SDG 11 in the smart city. Chapters 2–4 offer a detailed insight into the concepts of inclusiveness, safety, resilience, and sustainability and their meaning in the city space. Conceptual and empirical issues are stressed and elaborated. Part 2 of the volume engages with the question of why smart governance is a necessity for the attainment of the SDGs. The chapters forming this part of the book, Chapters 5–9, make the claim that to make the smart city truly smart it is necessary that the citizens' voices and so their feedback are heard, and that the citizens are involved in the making of the smart city. Part 3 of the volume focuses on the interconnected issues of safety, security, exclusion/inclusion, and well-being in the smart city. Chapters 10–12 elaborate on diverse aspects of these issues, including smart city planning and urban design, the emergence of ghettos and ways of preempting their expansion, cybersecurity issues pertinent to the smart city and its operation. Part 4 of the volume allows some space for sociological reflection on the smart city. Chapters 13 and 14 elaborate on diverse sociological aspects pertaining to the smart city and its development. Chapter 15 adds to this thread by demonstrating the value added of foresight exercises for sustainable smart city development. Chapter 16 concludes and suggests the way forward. Details on each chapter's content follow.

Following this introductory chapter, in Chapter 2, Raquel Pérez delHoyo, Anna Visvizi, and Higinio Mora, explore "Inclusiveness, safety, resilience, and sustainability in the smart city context," The increasing pace of urbanization and its geospatial distribution suggest that over the years to come cities and urban ecosystems will be exposed to a plethora of new challenges. The 2030 UN Agenda for Sustainable Development promotes the objective of cities being inclusive, safe, resilient, and sustainable. Yet, the crucial in this context, Goal 11, moves the responsibility for addressing issues and challenges related to the implementation of the SDGs to the gesture and responsibility of local governments. The problem is that not always local authorities are endowed with the necessary powers and resources to effectively implement respective policy goals and objectives. This only adds to the challenges cities faces today. This chapter queries the meaning of the key concepts defining the SDG 11 and suggests how ICT-enhanced applications and services may be useful in meeting the SDG 11.

In Chapter 3, titled "Sustainability and smart cities: What is in there? Bridging the theory and practice," Marco Tregua, Anna D'Auria, and Francesco Bifulco employ bibliometric analysis to examine how smart interventions are prioritized by city managers and combined in the practical realm of sustainable cities. To accomplish this goal, the authors perform a bibliometric analysis to depict the main elements describing sustainability in the smart city, with reference to people, technology, and policy. Successively, the discussion in the chapter moves from the evidence offered in the literature to practical experiences from cities at an advanced level of development as shown in the last 2019 IESE ranking. The authors argue that the interplay of elements such as people participation, technology-supported decision-making process, and governance involving an increasing number of actors are conducive to the attainment of the SDG 11 in the cities.

In Chapter 4, titled "Technology and the SDGs in the smart city context," Chrysaida-Aliki Papadopoulou highlights that the smart city represents a fundamental trend in urban development based on the exploitation of technology in order to support cities' sustainable development. Economic growth, social cohesion, and environmental protection are straightforwardly linked to the "smart" dimension of cities, implying the use of technology in almost all urban functions. Such pattern of urban development stimulates the accomplishment of sustainable development goals; especially the one concerning the promotion of safe, resilient, inclusive, and sustainable cities (SDG 11). Specific targets clarify this general perspective, highlighting issues such as standards of living, transportation services, climate change adaptation, etc. Each target is accompanied by an indicator measuring its performance. This chapter focuses on the role of smart technologies in achieving SDG 11. The key fundamental axes of smart cities are explored. Thereinafter, the focus shifts on the investigation of technologies supporting/upgrading city functions while their contribution to the accomplishment of SDG 11 is examined.

Chapter 5, titled "Governance and the SDGs in the smart city context," by Shahira Assem Abdel Razek, examines the term governance in context of the smart city. Whereas the term governance has been widely researched in a variety of fields (Visvizi, 2015), including the smart city context (Tomor et al., 2019), here the term governance is linked with urban space design and a city's sustainability. Against this backdrop, this chapter reviews the debate on urban development and governance to identify how these two strands of the debate address the issue of sustainability. A thorough discussion on how urban design can be conducive toward the attainment of the SDGs follows.

Chapter 6, titled "Reinterpreting governance and SDGs in smart cities: An ecosystems-based view," by Francesco Polese, Orlando Troisi, Mara Grimaldi, and Francesca Loia, explores the Italian municipality of Trento through the double lens of the smart city and smart urban ecosystem. The ecosystem-based view allows to identify different technologies, resources, and value co-creation practices that arise from actors' engagement with the (smart) city (space). If adequately harmonized by governance, these variables prove consequential for the co-development of sustainable innovation in the smart city. The findings and key arguments presented in the chapter are based on the case-study analysis, namely the case of TrentoSmartCity. This analysis allowed the authors to identify the main drivers of sustainable innovation in the smart city and to underline the harmonizing role of governance in this process. As the authors stress, the identification of the main governance steps for the (co-)development of innovation can help scholars and practitioners to shed light on the key role of managerial and governance skills (human component) in the successful exploitation of the innovation opportunities offered from technologies.

In Chapter 7, titled "Open data portals for urban sustainable development: People, policies, and technology," Christine Meschede and Tobias Siebenlist make a case for smart and sustainable cities as the key drives of sustainable urban development. The SDGs, as the authors argue, play a major role in this regard in that they can serve as an action plan for smart sustainable cities. With SDG 11, the importance of cities in context of sustainable development was reified. Moreover, and this makes this chapter stand out, the SDG 16 entails a target of access to information for everyone. Combining these goals, the availability of open urban data can promote a smart and sustainable development in cities. It can also supply viable data for measuring progress toward the SDGs. The authors base their discussion on the analysis of three national open data portals, including the cases of Australia, Germany, and South Africa.

In Chapter 8, titled "The potential of location-based social networks for participatory urban planning," Pablo Martí, Raquel Pérez-delHoyo, Almudena Nolasco-Cirugeda, Leticia Serrano-Estrada, and Clara García-Mayor explore the question of citizen participation in urban decision-making processes. As the authors argue citizen participation has proven to be of great value in view of aligning the population needs with urban planning proposals. However, in many cases, citizens frequently fail to be an active part of these planning processes. The SDG 11 offers urban planners the opportunity to reflect upon current strategies and tools to achieve more participatory and inclusive processes. In order to take a step forward in this direction, this paper studies the potential of five location-based social networks—Twitter, Instagram, Foursquare, Strava, and MapMyRun—to inform and improve urban planning processes. Within the context of smart cities, the main objective of this work is to present the advantages of incorporating user-generated information offered by these sources into decision-making processes to encourage smarter, participatory, and inclusive urban planning and management.

In Chapter 9, titled "The green turn: Smart cities, SDGs, and sustainability in the EU parties discourse," Cláudia Toriz Ramos examines the question of sustainability and the smart city through the lens of the SDG 11 and the green turn in European politics. The author argues that environmental issues became the leitmotiv in European politics and a major issue in recent electoral campaigns. Public awareness of the importance of the topic, notably resulting from the sustainable development goals (SDGs) agenda, reinforced this trend. However, the smart city issue cannot be tackled within a single policy area in the European Union (EU), as it requires a combination of approaches drawn from technology, energy and transport, and cohesion and regional development policies. The capacity to propose policies on sustainable cities requires from politicians a systemic approach to sustainability and, indeed, the ability to put forward ambitious plans for change. It is an assumption of this chapter that the EU can play a major role in shaping that change. It is also claimed that by engaging with this topic, the EU level parties can foster democratic processes in the EU. These are policy issues that touch the citizen's everyday life, and thus carry the potential to bring the citizens and the parties closer together. With reference to European political parties' discourse, especially at the time of the 2019 European Parliament elections, the chapter covers their main perspectives on environment, sustainability, and urban life.

Chapter 10, titled "Exploring the relationship between the smart-sustainable city, well-being, and urban planning: An analysis of current approaches in Europe," by Vasiliki Geropanta, Anna Karagianni, Sofia Mavroudi, and Panagiotis Parthenios, examines the cases of Barcelona, Copenhagen, and Vienna to make a case for the benefits of incorporating urban technologies into public space, As the authors argue this approach allows to promote successful urban development strategies that are aligned with the SDG 11.

Chapter 11, titled "A smart approach to identifying intra-urban disparities in neighborhood satisfaction: An example from Poland," by Piotr Maleszyk, offers an insight into the case of Lubin, a provincial, yet a thriving, university city, located in the eastern part of Poland. This chapter discusses the application of a smart tool for recognizing neighborhood satisfaction implemented in Lublin (Poland). The chapter also presents broader patterns in the "smart city" development in Poland, and then, specifically, in Lublin. In this way, the discussion in this chapter connects the smart city and the SDGs debates by focusing on and exploring in depth the question of intra-urban inequalities, their implications, and ways of addressing the municipality level.

Chapter 12, titled "Cybersecurity, sustainability and resilience capabilities of a smart city," by Roberto O. Andrade, Sang Guun Yoo, Luis Tello-Oquendo, and Iván Ortiz-Garcés, queries in which ways and to what extent emerging technologies may enhance the sustainability and resilience capacity of smart cities. As the authors point out, the prevalence of ICT-based solutions in the city space, increasingly makes smart cities vulnerable also to cybersecurity threats. However, increasingly the same technologies are susceptible to cybersecurity threats. This chapter examines the connection between cybersecurity and the sustainability and resilience capabilities of smart cities. To this end, this chapter maps the most common types of cybersecurity threats and their impact on smart city domains such as transportation, health, home, and others. These are examined and assessed by considering each layer of the smart city architecture proposed by the International Telecommunication Union (ITU). The discussion is embedded in the broader framework of the sustainable development goals (SDG11), in which the notion of cities' resilience and sustainability is stressed.

In Chapter 13, titled "Sociological reflections on smart cities and 'city times,'" Pablo de-Gracia-Soriano, Diana Jareño-Ruiz, María Jiménez-Delgado, and Antonio Jiménez-Delgado reflect on the ethical frameworks of smart cities. In this context the connection between the smart cities agenda and well-being of cities' inhabitants is discussed. Emphasis is placed on the concept of "city times," i.e., a term closely related to the lifestyles that take place in any social space, such as the city. A case is made for the conceptual suitability of "city times" for the analysis of the design of public policies in the smart city, especially in view of promoting well-being and healthy social life. Ultimately, the authors argue that that insights from sociology are of value in the process of developing smart cities and in the attainment of achievement of the SDGs.

In Chapter 14, titled "Smart city as the city of knowledge," Robert Laurini, Giuseppe Las Casas, and Beniamino Murgante view the smart city as the city of knowledge. As the authors argue, the 21st century is the knowledge society century. Knowledge sharing is thus seen as the key tool to enhance the human condition. This suggests the importance of education and culture as key factors necessary to empower the members of the society. It also suggests that knowledge is the driver of the economy and economic growth and development. Finally, it also suggests that knowledge is the key resource necessary for efficient governance. In this context, the key question is what are the implications of efficient knowledge sharing for the process of governance in the smart city? The authors examine this question by reference to the SDGs and their implementation in the European Union.

In Chapter 15, titled "Foresight exercises and the transformation of cities into smart cities," by Andrei Ştefan Neştian, and Ana Iolanda Voda, the value of foresight for the smart city is examined. As the authors argue, in the context of the 2030 Agenda for Sustainable Development, the definition of stakeholder community's vision about the future shape of the city and its transformation into a smart city is one of the key stages in smart city development. City foresight is viewed as the science of thinking about the future of cities. Urban communities tend to resort to foresight exercises as they help highlight and aggregate the needs of multiple stakeholders, suggest innovative solutions, and define a time horizon needed for developing realistic initiatives for city transformation. City foresight enables the exploration of alternative development paths and, therefore, provides practical tools for policy-makers to enhance their decision-making capabilities by using different scenario design exercises and exploratory hypotheses. The objective of this chapter is to shed light on the salience of foresight exercises in the transformation of cities into smart cities. For this purpose, we have systematized the foresight exercises and have made theoretical clarifications and classifications of forecast data, activities, and techniques used in city foresight, underlining the importance of foresight exercises for smart cities and their benefits. This systematization is useful for understanding the multiple alternatives, from which one can choose in order to define a foresight exercise, prior to putting forward practical initiatives transforming the city systems and public governance policies for a smart city. By means of conclusion, in Chapter 16, titled "Transforming the city from within: SDGs and their transformative dynamics in the smart city," Anna Visvizi and Raquel Perez del Hoyo outline how this volume contributes to the smart cities debate and in which ways it may be useful to the broader audience.

References

Ali, S., Hussain, T., Zhang, G., Nurunnabi, M., & Li, B. (2018). The implementation of sustainable development goals in "BRICS" countries. *Sustainability*, *10*, 2513. https://doi.org/10.3390/su10072513.

Anthopoulos, L. (Ed.). (2019). *Smart city emergence. Cases from around the world*. Amsterdam, Oxford, Cambridge: Elsevier.

Assefa, Y., Damme, W. V., Williams, O. D., & Hill, P. S. (2017). Successes and challenges of the millennium development goals in Ethiopia: Lessons for the sustainable development goals. *BMJ Global Health*, *2*(2), e000318. https://doi.org/10.1136/bmjgh-2017-000318.

Bali Swain, R., & Yang-Wallentin, F. (2020). Achieving sustainable development goals: Predicaments and strategies. *International Journal of Sustainable Development and World Ecology*, *27*(2), 96–106. https://doi.org/10.1080/13504509.2019.1692316.

Calzada, I., & Almirall, E. (2020). Data ecosystems for protecting European citizens' digital rights. *Transforming Government: People, Process and Policy*, *14*(2), 133–147. https://doi.org/10.1108/TG-03-2020-0047.

Cardullo, P., Di Feliciantonio, C., & Kitchinm, R. (Eds.). (2019). *The right to the smart city*. Bingley: Emerald Publishing. https://books.emeraldinsight.com/page/detail/The-Right-to-the-Smart-City/?k=9781787691407.

Carroll, J. (2020). One study, four cities: Information impact in neighborhood economic development. *Transforming Government: People, Process and Policy*. https://doi.org/10.1108/TG-07-2019-0070.

Christ, K. L., & Burritt, R. L. (2019). Implementation of sustainable development goals: The role for business academics. *Australian Journal of Management*, *44*(4), 571–593. https://doi.org/10.1177/0312896219870575.

Clarinval, A., Simonofski, A., Vanderose, B., & Dumas, B. (2020). Public displays and citizen participation: A systematic literature review and research agenda. *Transforming Government: People, Process and Policy*. https://doi.org/10.1108/TG-12-2019-0127.

Fukuda-Parr, S., Greenstein, J., & Stewart, D. (2013). How should MDG success and failure be judged: Faster progress or achieving the targets? *World Development*, 0305-750X. *41*, 19–30. https://doi.org/10.1016/j.worlddev.2012.06.014.

Gámez-García, D. C., Saldaña-Márquez, H., Gómez-Soberón, J. M., Corral-Higuera, R., & Arredondo-Rea, S. P. (2019). Life cycle assessment of residential streets from the perspective of favoring the human scale and reducing motorized traffic flow. From cradle to handover approach. *Sustainable Cities and Society*, 2210-6707. *44*, 332–342. https://doi.org/10.1016/j.scs.2018.10.018.

Gobin-Rahimbux, B., Cadersaib, Z., Chooramun, N., Gooda Sahib-Kaudeer, N., Heenaye-Mamode Khan, M., Cheerkoot-Jalim, S., et al. (2020). A systematic literature review on ICT architectures for smart Mauritian local council. *Transforming Government: People, Process and Policy*, *14*(2), 261–281. https://doi.org/10.1108/TG-07-2019-0062.

Gusmão Caiado, R. G., Leal Filho, W., Gonçalves Quelhas, O. L., de Mattos Nascimento, D. L., & Veigas Ávila, L. (2018). A literature-based review on potentials and constraints in the implementation of the sustainable development goals. *Journal of Cleaner Production*, 0959-6526. *198*, 1276–1288. https://doi.org/10.1016/j.jclepro.2018.07.102.

Kitchin, R., Cardullo, P., & Di Feliciantonio, C. (2019). Citizenship, justice and the right to the Smart City. In *The right to the smart city*. Bingley: Emerald Press.

Leal Filho, W. (2020). Viewpoint: Accelerating the implementation of the SDGs. *International Journal of Sustainability in Higher Education*, *21*(3), 507–511. https://doi.org/10.1108/IJSHE-01-2020-0011.

Lytras, M. D., & Visvizi, A. (2018). Who Uses Smart City Services and What to Make of It: Toward Interdisciplinary Smart Cities Research. *Sustainability*, *10*, 1998. https://doi.org/10.3390/su10061998.

Lytras, M. D., Visvizi, A., Chopdar, P. K., Sarirete, A., & Alhalabi, W. (2021). Information Management in Smart Cities: Turning end users' views into multi-item scale development, validation, and policy-making recommendations. *International Journal of Information Management*, *56*(2021), 102146. https://doi.org/10.1016/j.ijinfomgt.2020.102146.

Marvin, S., Luque-Ayala, A., & McFarlane, C. (2015). *Smart urbanism: Utopian vision or false dawn?*. Abingdon: Routledge.

Min, H., Soheil, K., & Kent, S. A. (Eds.). (2020). *Smart cities for technological and social innovation, case studies, current trends, and future steps*. Amsterdam, Oxford, Cambridge: Elsevier.

Miola, A., & Schiltz, F. (2019). Measuring sustainable development goals performance: How to monitor policy action in the 2030 agenda implementation? *Ecological Economics*, *164*, 106373. https://doi.org/10.1016/j.ecolecon.2019.106373.

Mora, L., & Deakin, M. (2019). *Untangling smart cities: From utopian dreams to innovation systems for a technology-enabled urban sustainability*. Amsterdam, Oxford, Cambridge: Elsevier.

Moreno-Ibarra, M., & Torres-Ruiz, M. (2019). Civic participation in smart cities: The role of social media. In A. Visvizi, & M. D. Lytras (Eds.), *Smart cities: Issues and challenges, mapping political, social and economic risks and threats*. Amsterdam, Oxford, Cambridge: Elsevier.

Moyer, J. D., & Hedden, S. (2020). Are we on the right path to achieve the sustainable development goals? *World Development*, 0305-750X. *127*, 104749. https://doi.org/10.1016/j.worlddev.2019.104749.

OECD. (2017). *Decentralisation and multi-level governance in Kazakhstan*. Paris: OECD Public Governance Reviews, OECD Publishing. https://doi.org/10.1787/9789264280700-en.

OECD. (2019). *Making decentralisation work: A handbook for policy-makers*. Paris: OECD Multi-Level Governance Studies, OECD Publishing. https://doi.org/10.1787/g2g9faa7-en.

OECD. (2020a). *A territorial approach to the sustainable development goals in Bonn, Germany*. OECD Regional Development Working Papers, No. 2020/07 Paris: OECD Publishing. https://doi.org/10.1787/bbf7e6b1-en.

OECD. (2020b). *The circular economy in cities and regions: Synthesis report*. Paris: OECD Urban Studies, OECD Publishing. https://doi.org/10.1787/10ac6ae4-en.

Ramirez-Rubio, O., Daher, C., Fanjul, G., et al. (2019). Urban health: An example of a "health in all policies" approach in the context of SDGs implementation. *Globalization and Health*, *15*, 87. https://doi.org/10.1186/s12992-019-0529-z.

Ruhil, R. (2015). Millennium development goals to sustainable development goals: Challenges in the health sector. *International Studies*, *52*(1–4), 118–135. https://doi.org/10.1177/0020881717725926.

Shmelev, S. (2019). *Sustainable cities reimagined: Multidimensional assessment and smart solutions*. London, New York: Routledge.

Tomor, Z., Meijer, A., Michels, A., & Geertman, S. (2019). Smart governance for sustainable cities: Findings from a systematic literature review. *Journal of Urban Technology*, *26*(4), 3–27. https://doi.org/10.1080/10630732.2019.1651178.

UNDP. (2010). *Evaluation of UNDO contribution to strengthening local governance*. New York: UNDP (United Nations Development Programme) Evaluation Office. https://www.oecd.org/derec/undp/47871446.pdf.

Vacca, J. (Ed.). (2020). *Solving urban infrastructure problems using smart city technologies: Handbook on planning, design, development, and regulation*. Amsterdam, Oxford, Cambridge: Elsevier.

van Soest, H. L., van Vuuren, D. P., Hilaire, J., Minx, J. C., Harmsen, M. J. H. M., Krey, V., et al. (2019). Analysing interactions among sustainable development goals with integrated assessment models. *Global Transitions*, *1*, 210–225. https://doi.org/10.1016/j.glt.2019.10.004.

Vandemoortele, J. (2018). From simple-minded MDGs to muddle-headed SDGs. *Development Studies Research*, *5*(1), 83–89. https://doi.org/10.1080/21665095.2018.1479647.

Visvizi, A. (2015). Safety, risk, governance and the Eurozone crisis: Rethinking the conceptual merits of 'global safety governance'. In P. Kłosińska-Dąbrowska (Ed.), *Essays on global safety governance: Challenges and solutions* (pp. 21–39). Warsaw: Centre for Europe, University of Warsaw, ASPRA-JR, ISBN:83-89547-24-4.

Visvizi, A., & Lytras, M. D. (Eds.). (2019a). *Smart cities: Issues and challenges: Mapping political, social and economic risks and threats* Elsevier, ISBN:9780128166390. https://www.elsevier.com/books/smart-cities-issues-and-challenges/lytras/978-0-12-816639-0.

Visvizi, A., & Lytras, M. D. (2019b). Reflecting on oikos and agora in smart cities context: Concluding remarks. In A. Visvizi, & M. D. Lytras (Eds.), *Smart cities: Issues and challenges: Mapping political, social and economic risks and threats* Elsevier, ISBN:9780128166390. https://www.elsevier.com/books/smart-cities-issues-and-challenges/lytras/978-0-12-816639-0.

Wu, S. M., Chen, T.-C., Wu, Y. J., & Lytras, M. (2018). Smart cities in Taiwan: A perspective on big data applications. *Sustainability*, *10*, 106. https://doi.org/10.3390/su10010106.

Yamagata, Y., & Yang, P. (Eds.). (2020). *Urban systems design: Creating sustainable smart cities in the internet of things era*. Amsterdam, Oxford, Cambridge: Elsevier.

Zait, A. (2017). Exploring the role of civilizational competences for smart cities' development. *Transforming Government: People, Process and Policy*, *11*(3), 377–392. https://doi.org/10.1108/TG-07-2016-0044.

Part I

Sustainability, SDGs, technology, and the city space: The smart city in the making

Chapter 2

Inclusiveness, safety, resilience, and sustainability in the smart city context

Raquel Pérez del Hoyo[a], **Anna Visvizi**[b,c], **and Higinio Mora**[a]
[a]*University of Alicante, Alicante, Spain,* [b]*SGH Warsaw School of Economics, Warsaw, Poland,* [c]*Effat University, Jeddah, Saudi Arabia*

1. Introduction

Cities face the challenge of accommodating up to 60% of the world's population in 2030 and up to 70% in 2050. The developing countries stand at the heart of the process of urbanization (UN, 2019a). The increasing pace of urbanization, or say, uncontrolled and voluntary migration to urban space, means that cities are exposed to a wide variety of challenges, including social exclusion, public health and public safety, mobility, resource supply, including clean water and energy, waste management, and others. To face these challenges and still be able to provide city inhabitants with dignified living conditions requires that a sustainable model of growth and development is adopted (OECD, 2012; Visvizi, Mazzucelli, & Lytras, 2017). The advent of sophisticated information and communication technology (ICT) and its application in urban/city space does not make the challenge any lesser, though. Actually, the advent of ICT, and so of smart city services, highlights that the value added of these and related solutions can only be pragmatized if cities are planned, designed, and managed effectively (Arasteh et al., 2016; Carroll, 2020; Ciasullo, Troisi, Grimaldi, & Leone, 2020; Lytras & Visvizi, 2020; Tjoa & Tjoa, 2016; Visvizi & Lytras, 2019). Sustainability and the Sustainable Development Goals (SDGs) Agenda, seen as a policy perspective and a policy objective, seem to define the trajectory of objectives and actions in this respect. This chapter queries this issue.

The United Nations (UN) *Transforming Our World: The 2030 Agenda for Sustainable Development* placed sustainability at the center of economic and social policies implemented at international, national, regional, and local levels (UN, 2015). Goal 11, which focuses on sustainable cities and communities, recognizes the essential role of cities as actors and catalysts of change. Specifically, Goal 11 states that cities, and so policies implemented at the city

Smart Cities and the UN SDGs. https://doi.org/10.1016/B978-0-323-85151-0.00002-6

level, should be inclusive, safe, resilient, and sustainable. In this view, efficient management and decision-making, inspired by these interrelated and interdependent principles, may serve as levers to increase cities' productivity and efficiency as regards boosting socioeconomic growth and development (Visvizi & Lytras, 2019). They can also serve as nodes' technological innovation (Polese, Tommasetti, Vesci, Carrubbo, & Troisi, 2016) and human capital development (Dziembała, 2019; Rybka-Iwańska & Serradell López, 2019).

The way the SGDs were framed, and in this context especially Goal 11, moved the responsibility for addressing issues and challenges related to the implementation of the SDGs to the gesture and responsibility of local governments. The problem is that not always local authorities are endowed with the necessary powers and resources to effectively implement respective policy goals and objectives. This only adds to the challenges cities face today.

Six years after the launch of the *Sustainable Development Goals Report* (UN, 2019b), some progress in these goals' implementation is visible; however, enormous challenges persist. In other words, even if 150 countries developed national urban plans to promote sustainable and inclusive growth, implementation is the bottleneck. Indeed, 9 out of 10 urban residents breathe polluted air, only half of the urban residents have convenient access to public transport, 1 out of 4 live in slums, and more than 2 billion people do not have waste collection services. There is therefore still a long way to go and, in this sense, this work is proposed, in the context of smart cities, to contribute to and support smart local government initiatives.

The objective of this chapter is twofold: to explore the meaning of the concepts determined by SDG 11, namely inclusiveness, safety, resilience, and sustainability; and to focus on how ICT-enhanced solutions applied in the smart city enable local governments in their pursuit to strengthen inclusiveness, safety, resilience, and sustainability. In the current context of smart cities, governments require efficient processes based on sophisticated technologies to strengthen urban systems and policies and, and in this way, to respond to citizens' needs. This paper provides an overview of how local governments employ ICT as a powerful ally in the process of making cities more inclusive, safe, resilient, and sustainable.

2. Rethinking the term "inclusive city"

The Global Campaign for Good Urban Governance of the United Nations (UN, 2000a) within the framework of the Habitat Agenda adopted as the central theme of its strategy "inclusion" and the vision of realizing the "inclusive city." The inclusive city was then defined as "a place where everyone, regardless of wealth, gender, age, race or religion, is enabled to participate productively and positively in the opportunities cities have to offer." In parallel, good urban governance was defined as one of the essential means to realize the inclusive city through "inclusive decision-making processes" capable of ensuring the highest probability of sustainability. Thus the idea of an inclusive city is global, even though the concept of inclusion is not, since it depends on the significance it has in each country, and it is therefore necessary to promote diverse governance policies and practices in each context.

With this new vision of the inclusive city adopted by the United Nations Centre for Human Settlements (Habitat), a substantial change is taking place in the way social inequality is understood, moving from isolation as urban poverty to being part of a much broader social concern that affects the city as a whole. This new vision, which is linked to the concept of citizenship, also implies a radical change in urban policy and a shift toward democratic governance, involving the public, private, and community sectors (Beall, 2000).

The aspects highlighted by the Campaign to be promoted for more inclusive cities were the following seven: sustainability, decentralization, equity, efficiency, transparency and accountability, civic engagement and citizenship, and security. Sustainability and security, including resilience in the security field, are aspects that the 2030 Agenda and this chapter consider separately. The rest of the issues are defined under the umbrella of the global concept of inclusiveness, with a distinction between the two areas. Decentralization, efficiency, transparency, and accountability are aspects that define good governance and are applied in its inclusive policies and practices. Equity, civil commitment, and citizenship are aspects directly linked to the concept of inclusion—as opposed to exclusion—of all groups in society.

From the perspective of inclusiveness, decentralization of authority and resources is proposed in order to deliver public services at all levels as efficiently and cost effectively as possible. Thus the concept of the inclusive city claims the autonomy of cities, an issue that has two important implications. On the one hand, cities need to take a greater role in the development of administrative, technical, and managerial tasks, and to perform these capacities in a transparent and accountable manner. On the other hand, cities have to be provided with greater resources as these must be sufficient to respond to the real needs of all citizens.

At the same time, the realization of the inclusive city is articulated on the basis of the twofold objective of achieving, on the one hand, full public participation and, on the other, equal access. One of the essential strategies for achieving inclusion is to promote the equitable participation of all civil society in civic life, and especially in decision-making processes, regardless of gender, age, race, or religion, and in particular of the most vulnerable groups. Another essential strategy focuses on the recognition of citizenship and civic engagement, and on the creation of a responsible ethic that encourages people to contribute to the common good. On the other hand, the protection of the rights of all people is a sine qua non for the realization of the inclusive city. Therefore another essential strategy to achieve this is to guarantee equal access both to decision-making processes, with the empowerment of the most vulnerable groups, and to the means to cover the basic needs for food, education, health, housing, work, etc., based on an equitable distribution of resources (da Cruz, Rode, & McQuarrie, 2019).

3. Challenges and opportunities of the "safe city"

The Habitat Agenda based its implementation on the promotion of two joint international campaigns. On the one hand, *The Global Campaign for Good Urban Governance* which, as we have seen, adopted the vision of the "inclusive city" (UN, 2000a); and, on the other, *The Global Campaign for Secure Tenure* focused on the right to housing (UN, 1999). Both campaigns introduced the issue of urban safety, the Secure Tenure Campaign referring to legal security of tenure for the most vulnerable groups, and the Governance Campaign addressing mediation and conflict resolution and environmental vulnerability issues. In addition, to support the implementation of the Habitat Agenda, the *Safer Cities Program* (1996) was launched to develop a strategy to prevent urban violence, in principle, in marginal communities where lack of safety has a disproportionate impact. Thus, as one of the universal objectives supported by the Habitat Agenda, the concept of safe human settlement was gradually given content, covering both the safety of the individual, his or her right to freedom and personal safety, and the safety of his or her environment, including environmental management and disaster prevention (Cozens, 2011; Kreimer, Arnold, & Carlin, 2003).

According to this vision, a "safe city" is a city that guarantees the safety of all citizens and their living environment, understanding that safety is a right of the people. Therefore safe cities are characterized by the care, vigilance, and protection of two fundamental aspects. Firstly, personal safety, through the prevention and control of crime and human conflicts, and also through the promotion of a civic culture based on solidarity, tolerance, respect, and acceptance of diversity. Another relevant aspect of personal safety is ensuring protection from persecution, forced evictions, as well as guaranteeing the safety of basic social services such as housing, health, and education. And secondly, environmental safety, through adequate environmental management and preparedness for natural and human disasters, promoting risk awareness and working to reduce vulnerability and in the preparation of management plans.

However, the vulnerability of each city is different, so addressing safety also involves different approaches. Moreover, cities today face new and changing safety challenges, whether in the social, infrastructure, or environmental fields (UN, 2016a). Digital safety is an issue of utmost urgency in cities (Elmaghraby & Losavio, 2014). Achieving a balance between the privacy and safety of citizens is a complex challenge that today, more than ever, has become evident from the management of the COVID-19 health crisis. The safety of cities is therefore a major challenge that, in many ways, has to do with their greater or lesser resilience.

4. The "resilient city": A new path to change the world

The concept of resilience has been applied in many areas of knowledge, from ecology to engineering, psychology, economics, urban planning, and information technology. The view from all these approaches is similar and follows the first definition introduced almost five decades ago (Holling, 1973). Holling (1973) linked the terms of resilience and stability. In this view, resilience is understood as a measure of the capacity of systems to absorb changes and to recover after a temporary disturbance. Resilience is at the core of all SDGs. It is also essential for sustainable progress. It is an important concept and policy objective in the debate on smart cities (Visvizi & Lytras, 2018).

The "resilient city" can therefore be defined as the city that is able to respond to threats of a very diverse nature that can be sudden changes, such as a natural disaster, a terrorist attack, an economic crisis or a health crisis like COVID-19, but which can also be more gradual transformations, such as dealing with inequality, an aging population, migration, resource depletion, waste management, or climate change. This suggests the broad scope of the term resilience and its multiple approaches (Vale, 2014).

On the other hand, resilience—like safety—has a preventive and anticipatory character in the urban context. It is necessary to be aware of and prepared with adequate management plans to face adverse situations while maintaining the basic functions and structures of a city. Therefore the concept of resilience also implies strengthening the transformative capacity of the city to anticipate future problems. This is why the resilient city has become one of the main objectives of global policy, being especially necessary to reduce vulnerability in the poorest countries. Furthermore, making cities more resilient is part of the strategy that allows them to prosper in the current context of a globalized world, where a rapid response to continuous change is essential, both to adapt the functioning of the city to new conditions and to recover its initial state and identity.

Knowledge of reality and awareness of problems strengthen the resilient city. Furthermore, resilience is circular and feeds back into lived experiences. But this alone is not enough; the scale of the problems facing cities requires good governance, accountable institutions, and the support of civil society. Consensus and collective, inclusive, and efficient management play a key role.

Resilient cities is a concept discussed in detail in the forum of the Organization for Economic Co-operation and Development (OECD) and in the OECD commissioned research (cf. Figueiredo, Honiden, & Schumann, 2018; OECD, 2018a, 2018b). Resilient cities are thus defined as cities that have the capacity to absorb, recover, and prepare for future shocks (economic, environmental, social, and institutional). However, resilient cities have also the capacity to promote sustainable development, well-being, and inclusive growth (cf. Visvizi & Lytras, 2018).

5. Toward a definition of "sustainable city" based on the social, economic, and environmental dimensions

Habitat I was the first United Nations Conference on Human Settlements to fully recognize the challenge of urbanization, advocating equity not only among nations but also among people (UN, 1976). A decade later, the first World Habitat

Day was celebrated (UN, 1985) with the aim of reflecting on the state of towns and cities, and on the right of all people to decent housing. In this context, the debate on the need to develop explicit strategies to guide the process of urbanization arises and the term sustainable development is defined for the first time as "that it meets the needs of the present without compromising the ability of future generations to meet their own needs" (UN, 1987). At the same time, the first population growth trends in urban areas were made public.

Shortly afterward, the United Nations Conference on Environment and Development (UN, 1992a) established Agenda 21 as a strategic global plan of action, calling on local governments to implement their own processes. Agenda 21 (UN, 1992b) identified the overall goal of promoting sustainable human settlements development and defined it as "to improve the social, economic and environmental quality of human settlements and the living and working environments of all people, in particular the urban and rural poor." In this way, the three universally accepted dimensions of sustainability were established: the social, the economic, and the environmental, whose priorities must balance cities in their development, not only thinking of the needs of the present but above all of future generations.

The holding of the second United Nations Conference on Human Settlements—Habitat II, also called the "The Cities Summit," led to the drafting of the Habitat Agenda (UN, 1996) which recognized the city as a "true engine of development." The Habitat Agenda gave continuity to Agenda 21 and its implementation, as mentioned above, focused on the promotion of international Campaigns for Urban Governance and Security of Tenure. Parallel to the development of these two campaigns, the new global alliance to reduce extreme poverty emerged and the eight Millennium Development Goals were established (UN, 2000b). These Millennium Development Goals will be followed by the implementation of the 2030 Agenda for Sustainable Development and its Sustainable Development Goals, which constitute today's ambitious global action plan for people, planet, and prosperity. The 2030 Agenda (UN, 2015) recognizes that eradicating poverty, as well as healing and securing our planet, is a prerequisite for sustainable development, and readopts the definition of sustainable development based on achieving a balance between the three dimensions of sustainability: social, economic, and environmental, without compromising future generations.

Only 1 year later and in the context of the 2030 Agenda, the third United Nations Conference on Housing and Sustainable Urban Development—Habitat III also placed greater emphasis on the concept of the sustainable city, resulting in the implementation of a New Urban Agenda from the previous Habitat Agenda. The objective of the New Urban Agenda (UN, 2016a) is none other than to serve as a means of addressing the great current challenge of planning and managing cities so that they become true engines of sustainable development, maintaining urban-rural links and the balance between the social, economic, and environmental dimensions of sustainable development. Thus the New Urban

Agenda, on the one hand, reaffirms the definition of sustainable development based on the three accepted dimensions of sustainability and, on the other hand, highlights the importance of long-term strategic planning of cities, emphasizing the opportunity that well-planned and managed urbanization represents for sustainable development.

No one doubts anymore the fundamental role that cities play in achieving sustainable development of the planet, nor the extremely important challenges of rapid urbanization, such as optimizing the use of resources, waste management, pollution reduction, eradication of poverty and slums, deterioration of infrastructure, equal access to basic services of transport, energy, water, sanitation, housing, etc. (Hoornweg & Freire, 2013).

According to data from *The Sustainable Development Goals Report* (UN, 2019b), although cities occupy only 3% of the earth's surface, they account for 60%–80% of energy consumption and 70% of carbon emissions. Virtually all low- and middle-income cities with more than 100,000 inhabitants do not meet the air quality guidelines, and the total amount of waste generated is estimated to double by 2050. An environmentally sustainable city is one that acts to reverse this situation with the utmost urgency, and is also an inclusive city because to reach agreement on acceptable levels of natural resource use, it is essential that all groups in society participate in the planning and environmental management processes. Urban planning also plays a key role in ensuring environmental sustainability. The consequences of poor planning, such as large, unnecessary expansion of cities, lower urban densities, the creation of slums, traffic problems, and consequent greenhouse gas emissions, have a major environmental impact.

On the other hand, the economically sustainable city must promote local economic growth and be competitive, financially viable, and favor quality employment. Promoting the participation of all citizens in the economic life of the city helps to achieve this, so the inclusive city is essential to improve economic sustainability. Furthermore, the economically sustainable city requires efficiency in the delivery of public services, which involves the contribution of the public, private, and community sectors.

Finally, social sustainability means making the city inclusive, which implies, as already mentioned, protecting the rights of all people, ensuring equal access to basic resources and services such as housing, and promoting participation in planning processes to address the needs of all, especially the most vulnerable groups (UN, 2000a). The socially sustainable city is one that promotes social cohesion at all levels, for example, by ensuring equitable distribution and convenient access for people to public open spaces—defined as spaces within 400 m of their residence. Only 31% of the population—with huge variations between cities—has convenient access to these spaces. Another important issue to avoid social segregation is access to sustainable public transport. Only 53% of the urban population has convenient access to public transportation—defined as living up to 500-m walking distance from a bus stop or similar, or 1000 m from a railway terminal or similar facility (UN, 2019b).

From all these considerations, the concept of the sustainable city can be deduced. A complete definition is provided by the United Nations in its explanation of Goal 11 on Sustainable Cities and Communities: "cities where all citizens live a decent quality of life, and form a part of the city's productive dynamic, creating shared prosperity and social stability without harming the environment" (UN, 2016b).

6. Using ICT toward developing inclusive, safe, resilient, and sustainable cities

In the current time of such rapid technological change, emerging technologies can be an important ally in implementing SDG 11 on cities. The potential of ICT can be useful in a wide range of areas, including those described in this section.

The effectiveness of e-government in the provision of public services depends on three main aspects: the adequacy of telecommunications infrastructure, the capacity of human resources to promote and use ICT, and the availability of online services and content, especially the latter aspect (UN, 2018). ICT makes it possible to improve both the accessibility of services and their effectiveness, and to provide them more quickly, reliably, and at a lower cost. They also enable governments to improve transparency and accountability to society (Pereira, Parycek, Falco, & Kleinhans, 2018).

Overall, ICT-enhanced applications and solutions in the smart city enable the delivery of increasingly equitable, personalized, and citizen-centric services (cf. Lytras & Visvizi, 2018; Lytras, Visvizi, Chopdar, Sarirete, & Alhalabi, 2021). In this way, these so-called smart city services play a key role in the development of the inclusive city. In this way, they also facilitate the delivery of e-services to the most vulnerable populations. It is becoming easier and less costly to access services through mobile devices and to get information to the most vulnerable groups through websites (Albesher & Stone, 2016). A good example of this is the Republic of Korea, which has become a paradigm of e-government performance, with the implementation of a high-quality information system for citizens (Nam, Oh, Kim, Goo, & Khan, 2016). In general, mobile technology has enabled more users to participate in the design and delivery of services. The widespread use of SMS and social networking combined with crowdsourcing approaches has been a key facilitator for social change (Gigler et al., 2014). However, the services provided are not always considered useful by users (Lytras & Visvizi, 2018) and e-participation processes have not always allowed citizens to participate in decision-making, limiting themselves to collecting their opinions and suggestions (Cardoso Sampaio, 2016).

Another field in which ICT has great potential in cities is that of safety. A safe city should have smart safety systems to address many different aspects: health care; traffic routes; surveillance, search, detection, and identification; crisis management to support decision-making, early warning, monitoring and forecasting of emergency and environmental situation; centralization of emergency, police and rescue services; secure Internet connection and data protection; data processing centers, among others (Lacinák & Ristvej, 2017).

In the field of personal safety, technologies such as CCTV cameras, urban supervision centers, or platforms for sharing information not only contribute to crime prevention (Chiodi, 2016) or to the management of public transport (Sutar, Koul, & Suryavanshi, 2016) but also to understanding its causes (Cardone et al., 2013). In the field of environmental security, ICT is essential for the rapid and efficient management of emergencies and natural or man-made disasters. It is a great support in the prevention and reduction of risks, as well as in the response to disasters and in the subsequent recovery phase. Timely coordination and information sharing is critical and there is an increasing reliance on ICT in crisis management. For example, Governments are increasing the use of Geographic Information Systems (GIS) and open geospatial data (Tomaszewski, 2020) to strengthen the resilient city. GIS technology can facilitate disaster risk assessment and management based on geospatial data analysis, both in decision-making aimed at risk reduction and prediction and in assessing damage and mitigating it at a later stage.

However, digital transformation requires planning and continuous updating to ensure safety and not to deviate from the goal of sustainable development (UN, 2018). ICT is a key factor in social and economic development in an interconnected world. However, in this context of global connectivity, the implementation of ICT is also vulnerable to cyberattacks. Cybersecurity is increasingly relevant to ensure the continuity of online public services, as well as to protect data and individual privacy (Khatoun & Zeadally, 2017). Greater awareness of cybersecurity is therefore required, since the inappropriate use of ICT can put the safety of States, businesses, and societies in general at risk.

Despite this, technologies such as Big Data, smart applications, and analytics offer endless possibilities. Cities generate a great deal of data, but this data has to be processed and analyzed to make knowledge accessible on demand. This is done through automated smart applications and analytics that reveal trends in users, their preferences, population movements, their use of services, transport, and public spaces. Knowing this information allows managers to improve these services and make more appropriate decisions. It is thus possible to say that there is a new science of data (OECD, 2017). Moreover, making the data available to different groups enables the possibilities of action to be expanded. Open data are being used by several countries with important benefits, for example, the use of geospatial data to improve urban planning in Denmark (Turner, Odgaard, Bøcher, Dalgaard, & Svenning, 2014), or historical air quality data to mitigate pollution in Mexico (Moraga & Rabiei Hosseinabad, 2017).

If artificial intelligence is added to data technology, a wide range of processes can be optimized and automated, such as climate change forecasting, traffic and transport management, crime prevention, the fight against disease, or communication with citizens. For example, artificial intelligence can predict environmental changes from the analysis of large data and geolocated images and provide evidence to make decisions in the direction of SDGs, as in the case of Congo (Christensen & Jokar Arsanjani, 2020).

The data, which can be collected through sensors on objects connected through the Internet of things, can be processed centrally through cloud computing, or at the collection point through edge computing. This enables, for example, the creation of increasingly smart and environmentally friendly homes and urban environments, facilitating sustainable energy use, proper air quality management, waste collection, or infrastructure maintenance. This real-time data from sensors that can monitor energy consumption, air pollution, or traffic can also be combined with data retrieved from social media or mobile devices to enrich knowledge. The real-time analysis of all this data and high-performance computing enables us to act quickly in the face of potential threats. A good example of an open platform for the collection, processing, and distribution of real-time data generated in a city is the one in Singapore (Kloeckl, Senn, & Ratti, 2012).

If virtual reality is added to the above technologies, the user experience can be substantially improved by generating a digital world that provides context. If the added technology is augmented reality, then it is the real world that is enriched by interacting in real time with added digital objects, which improves the visualization of possible scenarios. These technologies are useful for urban planning, public participation, infrastructure management, transport, tourism, or safety services. For example, augmented reality has been very beneficial in improving the experience of people with disabilities in certain environments (Rashid, Melià-Seguí, Pous, & Peig, 2017).

Finally, also the blockchain and distributed ledger technology offer interesting possibilities for the city administration (Batubara, Ubacht, & Janssen, 2018). These technologies enable information to be stored in a distributed manner among multiple actors. With the decentralization of the data, more security and privacy is achieved—although also more complex—which enables problems of access and control of the information to be solved. Therefore these technologies are very useful to establish the framework of relationship between governments and citizens for the exchange of data. For example, they are useful for the payment of services and taxes, to manage registers, request certificates or personal documents, as well as other documents or licenses, etc. A good example of implementation is the one related to the application for planning permission (Mora, Pérez-delHoyo, Sirvent, & Gilart-Iglesias, 2019).

7. Conclusions

Cities and urban ecosystems are exposed to a mounting variety of risks and challenges (Visvizi & Lytras, 2019). Whereas the objective of making cities inclusive, safe, resilient, and sustainable has been ingrained in the SDGs, the responsibility for addressing issues and challenges related to the implementation of the SDGs has been left in the hands of local governments. The latter frequently lack both the powers and resources to implement strategies necessary to attain inclusion, safety, resilience, and sustainability. This chapter examined the meaning of the key concepts defining SDG 11 and suggested ways in which ICT-enhanced applications and services can be useful in implementing these key concepts and policy imperatives.

References

Albesher, A. S., & Stone, R. T. (2016). Current state of m-government research: Identifying future research opportunities. *International Journal of Electronic Governance*, *8*(2), 119–139. https://doi.org/10.1504/IJEG.2016.078118.

Arasteh, H., Hosseinnezhad, V., Loia, V., Tommasetti, A., Troisi, O., Shafie-Khah, M., et al. (2016). Iot-based smart cities: A survey. In *2016 IEEE 16th international conference on environment and electrical engineering (EEEIC), Florence, 2016* (pp. 1–6). https://doi.org/10.1109/EEEIC.2016.7555867.

Batubara, F. R., Ubacht, J., & Janssen, M. (2018). Challenges of blockchain technology adoption for e-government: a systematic literature review. In *Proceedings of the 19th annual international conference on digital government research: Governance in the data age* (pp. 1–9). https://doi.org/10.1145/3209281.3209317.

Beall, J. (2000). From the culture of poverty to inclusive cities: Re-framing urban policy and politics. *Journal of International Development*, *12*(6), 843–856. https://doi.org/10.1002/1099-1328(200008)12:6<843::AID-JID713>3.0.CO;2-G.

Cardone, G., Foschini, L., Bellavista, P., Corradi, A., Borcea, C., Talasila, M., et al. (2013). Fostering participaction in smart cities: A geo-social crowdsensing platform. *IEEE Communications Magazine*, *51*(6), 112–119. https://doi.org/10.1109/MCOM.2013.6525603.

Cardoso Sampaio, R. (2016). e-Participatory budgeting as an initiative of e-requests: Prospecting for leading cases and reflections on e-participation. *Revista Brasileira de Administração Pública*, *50*(6). https://doi.org/10.1590/0034-7612152210.

Carroll, J. (2020). One study, four cities: Information impact in neighborhood economic development. *Transforming Government: People, Process and Policy*. https://doi.org/10.1108/TG-07-2019-0070.

Chiodi, S. I. (2016). Crime prevention through urban design and planning in the smart city era. *Journal of Place Management and Development*, *9*(2), 137–152. https://doi.org/10.1108/JPMD-09-2015-0037.

Christensen, M., & Jokar Arsanjani, J. (2020). Stimulating implementation of sustainable development goals and conservation action: Predicting future land use/cover change in Virunga National Park, Congo. *Sustainability*, *12*(4), 1570. https://doi.org/10.3390/su12041570.

Ciasullo, M. V., Troisi, O., Grimaldi, M., & Leone, D. (2020). Multi-level governance for sustainable innovation in smart communities: An ecosystems approach. *The International Entrepreneurship and Management Journal*, *16*, 1167–1195. https://doi.org/10.1007/s11365-020-00641-6.

Cozens, P. M. (2011). Urban planning and environmental criminology: Towards a new perspective for safer cities. *Planning Practice and Research*, *26*(4), 481–508. https://doi.org/10.1080/02697459.2011.582357.

da Cruz, N. F., Rode, P., & McQuarrie, M. (2019). New urban governance: A review of current themes and future priorities. *Journal of Urban Affairs*, *41*(1), 1–19. https://doi.org/10.1080/07352166.2018.1499416.

Dziembała, M. (2019). Smart city as a steering center of the region's sustainable development and competitiveness. In A. Visvizi, & M. D. Lytras (Eds.), *Smart cities: Issues and challenges* (pp. 149–169). Elsevier, ISBN:9780128166390. https://doi.org/10.1016/B978-0-12-816639-0.00009-0.

Elmaghraby, A. S., & Losavio, M. M. (2014). Cyber security challenges in smart cities: Safety, security and privacy. *Journal of Advanced Research*, *5*(4), 491–497. https://doi.org/10.1016/j.jare.2014.02.006.

Figueiredo, L., Honiden, T., & Schumann, A. (2018). *Indicators for resilient cities*. Paris: OECD Regional Development working papers, no. 2018/02, OECD Publishing. https://doi.org/10.1787/6f1f6065-en.

Gigler, B. S., Custer, S., Bailur, S., Dodds, E., Asad, S., & Gagieva-Petrova, E. (2014). Closing the feedback loop: Can technology amplify citizen voices? In *Closing the feedback loop: Can technology bridge the accountability gap?* (pp. 211–277). The World Bank. https://doi.org/10.1596/978-1-4648-0191-4.

Holling, C. S. (1973). Resilience and stability of ecological systems. *Annual Review of Ecology and Systematics*, *4*(1), 1–23. https://doi.org/10.1146/annurev.es.04.110173.000245.

Hoornweg, D., & Freire, M. (2013). *Building sustainability in an urbanizing world: A partnership report*. The World Bank. Retrieved from: http://documents.worldbank.org/curated/en/622651468320375543/pdf/801060REVISED00170Partnership0Paper.pdf.

Khatoun, R., & Zeadally, S. (2017). Cybersecurity and privacy solutions in smart cities. *IEEE Communications Magazine*, *55*(3), 51–59. https://doi.org/10.1109/MCOM.2017.1600297CM.

Kloeckl, K., Senn, O., & Ratti, C. (2012). Enabling the real-time city: LIVE Singapore! *Journal of Urban Technology*, *19*(2), 89–112. https://doi.org/10.1080/10630732.2012.698068.

Kreimer, A., Arnold, M., & Carlin, A. (Eds.). (2003). *Building safer cities: The future of disaster risk* The World Bank. https://doi.org/10.1596/0-8213-5497-3.

Lacinák, M., & Ristvej, J. (2017). Smart city, safety and security. *Procedia Engineering*, *192*, 522–527. https://doi.org/10.1016/j.proeng.2017.06.090.

Lytras, M. D., & Visvizi, A. (2018). Who uses smart city services and what to make of it: Toward interdisciplinary smart cities research. *Sustainability*, *10*(6), 1998. https://doi.org/10.3390/su10061998.

Lytras, M. D., & Visvizi, A. (2020). Information management as a dual-purpose process in the smart city: Collecting, managing and utilizing information. *International Journal of Information Management*. https://doi.org/10.1016/j.ijinfomgt.2020.102224.

Lytras, M. D., Visvizi, A., Chopdar, P. K., Sarirete, A., & Alhalabi, W. (2021). Information Management in Smart Cities: Turning end users' views into multi-item scale development, validation, and policy-making recommendations. *International Journal of Information Management*, *56*, 102146. https://doi.org/10.1016/j.ijinfomgt.2020.102146.

Mora, H., Pérez-delHoyo, R., Sirvent, R. M., & Gilart-Iglesias, V. (2019). Management city model based on blockchain and smart contracts technology. In *The international research & innovation forum* (pp. 305–313). Cham: Springer. https://doi.org/10.1007/978-3-030-30809-4_28.

Moraga, R., & Rabiei Hosseinabad, E. (2017). A system dynamics approach in air pollution mitigation of metropolitan areas with sustainable development perspective: A case study of Mexico City. *Journal of Applied Environmental and Biological Sciences*, *7*(12), 164–174.

Nam, K., Oh, S. W., Kim, S. K., Goo, J., & Khan, M. S. (2016). Dynamics of enterprise architecture in the Korean public sector: Transformational change vs. transactional change. *Sustainability*, *8*(11), 1074. https://doi.org/10.3390/su8111074.

OECD. (2012). Organisation for economic co-operation and development. In *OECD environmental outlook to 2050. The consequences of inaction* OECD Publishing. https://doi.org/10.1787/9789264122246-en.

OECD. (2017). Organisation for economic co-operation and development. In *Development co-operation report 2017: Data for development* OECD Publishing. https://doi.org/10.1787/dcr-2017-en.

OECD. (2018a). *Building resilient cities: An assessment of disaster risk management policies in Southeast Asia*. Paris: OECD Green Growth Studies, OECD Publishing. https://doi.org/10.1787/9789264305397-en.

OECD. (2018b). Promoting sustainable, resilient and inclusive cities in Myanmar. In *Case studies on leaving no one behind: A companion volume to the development co-operation report 2018*. Paris: OECD Publishing. https://doi.org/10.1787/9789264309333-28-en.

Pereira, G. V., Parycek, P., Falco, E., & Kleinhans, R. (2018). Smart governance in the context of smart cities: A literature review. *Information Polity, 23*(2), 143–162. https://doi.org/10.3233/IP-170067.

Polese, F., Tommasetti, A., Vesci, M., Carrubbo, L., & Troisi, O. (2016). Decision-making in smart service systems: A viable systems approach contribution to service science advances. In T. Borangiu, M. Dragoicea, & H. Nóvoa (Eds.), *Vol. 247. Exploring services science. IESS 2016. Lecture notes in business information processing*. Cham: Springer. https://doi.org/10.1007/978-3-319-32689-4_1.

Rashid, Z., Melià-Seguí, J., Pous, R., & Peig, E. (2017). Using augmented reality and internet of things to improve accessibility of people with motor disabilities in the context of smart cities. *Future Generation Computer Systems, 76*, 248–261. https://doi.org/10.1016/j.future.2016.11.030.

Rybka-Iwańska, K., & Serradell López, E. (2019). Smart cities and the search for global talent. In A. Visvizi, & M. D. Lytras (Eds.), *Smart cities: Issues and challenges* (pp. 171–184). Elsevier, ISBN:9780128166390. https://doi.org/10.1016/B978-0-12-816639-0.00010-7.

Sutar, S. H., Koul, R., & Suryavanshi, R. (2016). Integration of Smart Phone and IOT for development of smart public transportation system. In *2016 International conference on internet of things and applications (IOTA)* (pp. 73–78). IEEE. https://doi.org/10.1109/IOTA.2016.7562698.

Tjoa, A. M., & Tjoa, S. (2016). The role of ICT to achieve the UN sustainable development goals (SDG). In *ICT for promoting human development and protecting the environment. WITFOR 2016* (pp. 3–13). Cham: Springer. https://doi.org/10.1007/978-3-319-44447-5_1.

Tomaszewski, B. (2020). *Geographic Information Systems (GIS) for disaster management* (2nd). New York: Routledge. https://doi.org/10.4324/9781351034869.

Turner, K. G., Odgaard, M. V., Bøcher, P. K., Dalgaard, T., & Svenning, J. C. (2014). Bundling ecosystem services in Denmark: Trade-offs and synergies in a cultural landscape. *Landscape and Urban Planning, 125*, 89–104. https://doi.org/10.1016/j.landurbplan.2014.02.007.

UN. (1976). *United Nations-Habitat. The Vancouver declaration on human settlements*. Retrieved from: https://mirror.unhabitat.org/downloads/docs/The_Vancouver_Declaration.pdf.

UN. (1985). *United Nations General Assembly. A/RES/40/202. Report of the Commission on Human Settlements*. Retrieved from: https://undocs.org/A/RES/40/202.

UN. (1987). United Nations General Assembly. A/42/427. The urban challenge. In *Report of the World Commission on Environment and Development: Our common future*. Retrieved from: https://sustainabledevelopment.un.org/content/documents/5987our-common-future.pdf (chapter 9).

UN. (1992a). *United Nations General Assembly. A/CONF.151/26. Vol. 1.* Rio Declaration on Environment and Development. Retrieved from: https://www.un.org/en/development/desa/population/migration/generalassembly/docs/globalcompact/A_CONF.151_26_Vol.I_Declaration.pdf.

UN. (1992b). United Nations Commission on Sustainable Development. Promoting sustainable human settlement development. In *Agenda 21*. Retrieved from: https://sustainabledevelopment.un.org/content/documents/Agenda21.pdf (chapter 7).

UN. (1996). *United Nations conference on human settlements (Habitat II). A/CONF.165/14*. Retrieved from: https://undocs.org/A/CONF.165/14.

UN. (1999). *The global campaign for secure tenure: Implementing the habitat agenda—Adequate shelter for all*. United Nations. Centre for Human Settlements (Habitat). Retrieved from: http://www.vasilievaa.narod.ru/mu/unhabitatmoscow/tenure.asp.htm.

UN. (2000a). United Nations. Centre for Human Settlements (Habitat). The global campaign for good urban governance. *Environment and Urbanization*, *12*(1), 197–202. https://doi.org/10.1177/095624780001200114.

UN. (2000b). *United Nations General Assembly. A/RES/55/2. Millennium declaration*. Retrieved from: https://www.un.org/en/development/desa/population/migration/generalassembly/docs/globalcompact/A_RES_55_2.pdf.

UN. (2015). *Transforming our world: The 2030 agenda for sustainable development. A/RES/70/1*. United Nations. Retrieved from: https://sustainabledevelopment.un.org/content/documents/21252030%20Agenda%20for%20Sustainable%20Development%20web.pdf.

UN. (2016a). *New urban agenda*. United Nations-Habitat III. Retrieved from: http://habitat3.org/wp-content/uploads/NUA-English.pdf.

UN. (2016b). *Sustainable cities: Why they matter*. United Nations. Retrieved from: https://www.un.org/sustainabledevelopment/wp-content/uploads/2016/08/11.pdf.

UN. (2018). *Department of Economic and Social Affairs. United Nations E-Government Survey 2018: Gearing E-Government to support transformation towards sustainable and resilient societies*. United Nations. Retrieved from: https://www.unescap.org/sites/default/files/E-Government%20Survey%202018_FINAL.pdf.

UN. (2019a). *Department of Economic and Social Affairs. World Urbanization Prospects 2018. Highlights*. United Nations. Retrieved from: https://population.un.org/wup/Publications/Files/WUP2018-Highlights.pdf.

UN. (2019b). *The sustainable development goals report 2019*. United Nations. Retrieved from: https://unstats.un.org/sdgs/report/2019/The-Sustainable-Development-Goals-Report-2019.pdf.

Vale, L. J. (2014). The politics of resilient cities: Whose resilience and whose city? *Building Research and Information*, *42*(2), 191–201. https://doi.org/10.1080/09613218.2014.850602.

Visvizi, A., & Lytras, M. (2018). Rescaling and refocusing smart cities research: From mega cities to smart villages. *Journal of Science and Technology Policy Management*. https://doi.org/10.1108/JSTPM-02-2018-0020.

Visvizi, A., & Lytras, M. (Eds.). (2019). *Smart cities: Issues and challenges: Mapping political, social and economic risks and threats* Elsevier, ISBN: 9780128166390 https://www.elsevier.com/books/smart-cities-issues-and-challenges/lytras/978-0-12-816639-0.

Visvizi, A., Mazzucelli, C. G., & Lytras, M. (2017). Irregular migratory flows: Towards an ICTs' enabled integrated framework for resilient urban systems. *Journal of Science and Technology Policy Management*, *8*(2), 227–242. https://doi.org/10.1108/JSTPM-05-2017-0020.

Chapter 3

Sustainability in smart cities: Merging theory and practice

Marco Tregua, Anna D'Auria, and Francesco Bifulco
University of Naples Federico II, Naples, Italy

1. Introduction and literature overview

Sustainability and management of urban contexts are very hot topics in the contemporary academic debate, in which the two topics are often joined in relationship with the field of smart cities. In line with this, the smart city can be considered a starting point for developing policies for long-term sustainability for urban areas aimed at putting into practice the global axiom of "Think globally, act locally" (Gilderbloom, Squires, Riggs, & Čapek, 2017; Rudie & Wonham, 1991; Visvizi & Lytras, 2018) and in line with Sustainable Development Goals (SDGs) proposed by the United Nations and recalled by other scholars in previous years (De Guimarães, Severo, Júnior, Da Costa, & Salmoria, 2020; Klopp & Petretta, 2017; Visvizi, Lytras, Damiani, & Mathkour, 2018).

In recent years, the conceptualization of the smart city has generated a large number of studies by scholars, central or local institutions, and industry players involved in smart projects (Albino, Berardi, & Dangelico, 2015; Bakıcı, Almirall, & Wareham, 2013; Bifulco, Tregua, Amitrano, & D'Auria, 2016; Borda & Bowen, 2019; Nam & Pardo, 2011). Furthermore, an increasing number of smart transformation projects and initiatives have been proposed and carried out. However, the configuration of the smart urban context is still hard to shape, and the proposed models remain complex and diversified, as several differences emerge when one is comparing features of local contexts influencing smart projects in different geographical areas (Neirotti, De Marco, Cagliano, Mangano, & Scorrano, 2014).

In general, the new challenges led cities to begin the process of smart transformation through the introduction of technology in every aspect of the urban environment, so as to offer a better quality of life for stakeholders (Lytras, Visvizi, Chopdar, Sarirete, & Alhalabi, 2021; Mora, Gilart-Iglesias, Pérez-del Hoyo, & Andújar-Montoya, 2017; Visvizi & Lytras, 2019). As time went by, scholars, institutions, large companies, and corporations proposed several definitions or labels (i.e., digital, intelligent, ubiquitous, wired, hybrid, information,

Smart Cities and the UN SDGs. https://doi.org/10.1016/B978-0-323-85151-0.00003-8

creative, learning, humane, knowledge, and smart) aimed at describing the renewed configurations acquired by the local contexts under the engine of globalization (D'Auria, Tregua, & Vallejo-Martos, 2018; Nam & Pardo, 2011). In most of these contexts, ICTs and a sustainable approach are drivers to develop a more efficient urban context; therefore the long-lasting and more considered definitions, probably because of their broader approach, are "digital city" and "smart city" (e.g., Dameri, 2017; Ishida, 2017).

Local administrations and firms must consider each aspect of everyday life in shaping a smart urban context, as clearly illustrated by the model of Giffinger, Fertner, Kramar, and Meijers (2007) with the different elements embedded and classified in six macro-categories defined as "characteristics" or "drivers," namely, smart mobility, smart environment, smart living, smart people, smart governance, and smart economy. According to Giffinger et al. (2007), the integration of the six areas through traditional and modern Information and Communication infrastructures could lead to the aimed sustainable growth.

A scan of scientific contributions and official reports reveals that the definition of sustainability has been used mainly to underline the relevance of the preservation of natural resources, social equity, and economic development in search of a systemic vision with elements that contribute to achieving common goals (De Jong, Joss, Schraven, Zhan, & Weijnen, 2015; Kumar, 2020). Indeed, due to rapid population growth, urban contexts are now facing issues such as the overexploitation of resources, an increase in pollution, global overheating, and an inadequate number of services for citizens. These aspects are significantly affecting people's quality of life; therefore public administrations—both local and central—are pursuing sustainable development goals to overcome these criticalities (De Guimarães et al., 2020; Klopp & Petretta, 2017).

Several scholars have stressed the relevance of adopting a sustainable approach within territories' development and called for further research (Ghisetti, 2018). For example, Schilling and Logan (2008) underlined how sustainable strategies in cities can create new urban scenarios, with significant benefits for the local communities, while Yigitcanlar and Lönnqvist (2013) focused on the role that knowledge plays in city design and on strategies concerning the different dimensions of sustainable development. Furthermore, several scholars have stressed the role of the approach to and challenges of sustainability and new technologies in achieving sustainable goals (Bifulco, D'Auria, Amitrano, & Tregua, 2018; Hollands, 2015; Nižetić, Djilali, Papadopoulos, & Rodrigues, 2019; Soe, 2020).

It is worth noting that sustainability in the smart city refers not only to environmental issues; in this regard, scholars have focused on the economic domain of sustainability to analyze existing tools to favor, activate, and evaluate sustainable development and to create new ones (Brorström, Argento, Grossi, Thomasson, & Almqvist, 2018), also merging multidisciplinary approaches (e.g., Lytras & Visvizi, 2018). For example, Tanguay, Rajaonson, Lefebvre, and Lanoie (2010) performed an analysis to identify suitable indicators to apply in smart cities, while Komninos (2013) delineated some instruments to support the implementation of enabling platforms for the development of innovative city services. Another interesting contribution has been provided by Batagan (2011), who proposed a set of indicators for the economic and social development of smart cities, emphasizing that modern technologies make the urban context more efficient and play a relevant role in modernization and sustainable growth.

To identify the results that have already been achieved through smart projects and, in particular, to investigate how to attain a smart city, this chapter analyzes both theoretical and practical advances through a double-step analysis. The remainder of the discussion in this chapter is structured as follows: The section on methodology depicts the research design, the cities chosen for the analysis, and the sources used; then the findings are articulated on the three pillars, namely, people, technology, and policy, with each section combining results from the bibliometric analysis and evidence from cities. Finally, the section on implications depicts the outcomes for scholars, practitioners, and policymakers. The chapter ends with conclusions and suggestions for further research.

2. Methodology and research design

To investigate the role of sustainability in smart interventions and the evolution of the new shapes of urbanism linked to the sustainable development perspective, the authors designed a research process based on a double methodology, namely, a bibliometric and content analysis. First, the literature review provides evidence of only a limited number of contributions regarding the sustainable city; thus an additional step of analysis can help achieve better insights. Moreover, the context of smart interventions is permeated by institutional actors, so the opportunity to consider their perspectives was considered interesting. This led the authors to perform a multiple case study analysis too. Combining a previous research step with other methods is suggested (Kohlbacher, 2006) when data material (like reports, in this case) emerges as interesting sources of information. To sum up, our research approach is based on highlighting the key advances in the literature and in practice with reference to people, technology, and policy in smart cities, in order to understand how these elements may favor a sustainable outcome. Therefore the analysis offered in the next section is based on the combination of these three pillars, due to the role they play in smart cities development. We framed our research approach using a multilevel model, consisting of micro-, meso-, and macro-levels, in line with previous studies on city transformation (e.g., Johnson, 2012). These three levels may be combined with, respectively, people, technology, and policy, according to Jayashree, Hamza, El Barachi, and Gholami (2019), because (a) people act at the microlevel as addressees of smart services and as contributors of bottom-up ideas to stimulate change; (b) technology is an instrument operating at the meso-level because it links citizens with other actors at the political level, and favoring service provision, data exchange, data collection, as well as innovation based on data analysis; and (c) policy represents the macro-level because the contributions from users—i.e., citizens—are addressed from the local, national, and supranational entities, and because smart services are planned and provided at an institutional level. Additionally, the three levels are not considered as stand-alone; their interplay will be considered.

As it concerns the sources and the selection criteria, the dataset for the bibliometric analysis has been built by considering contributions from all the different branches of science, as several fields of research are directly linked to the smart city and sustainability. Then key elements have been used to perform the analysis, namely, people, technology, and policies, identifying micro-, meso-, and macro-levels. The metadata of contributions have been collected from the database of the Web of Science—Core Collection, as it is considered the most reliable database for performing such an investigation (Bremholm, 2004). The author selected the data through three queries shaping a set of items (papers published in journals until 2020, with the first result dating back to 2011) to perform the analysis through Bibliometrix (Aria & Cuccurullo, 2017). The investigation focused on the connection emerging among the themes in a thematic network.

Finally, the cases supported the evidence from the above step and were chosen because of their relevance; indeed, the IESE Cities in Motion 2019 ranking offered the opportunity to select the most relevant cases throughout the world. The top 20 cities were considered as a result and were analyzed as cases through multiple sources of data (websites, official reports, articles published online, and so on).

3. Findings: Smart cities challenging sustainability

This section consists of three main elements shaping scholars' discussion of smart urban contexts, namely, people, technology, and policy. As highlighted above, the debate over sustainability in smart cities has attracted the attention of several scholars from various fields of science. Indeed, about 700 journal articles shaped the entire dataset of this research. Among all these contributions, only 2.5% combine all three considered elements and favor the infusion of sustainability in smart cities.

3.1 People: Theoretical evidence

The first dataset consists of the contributions in scientific journals debating over people, sustainability, and smart cities. The results are categorized into two parts, namely, one considering the qualitative outcome aimed at people, and the other dealing with the quantitative side, viz., population density and small-sized communities. These parts show a wide range of results among all those from the bibliometric analysis, as in the thematic network in Fig. 1.

The first dataset consists of the relations among people, sustainability, and the smart city; these deal with (a) people's participation in smart governance

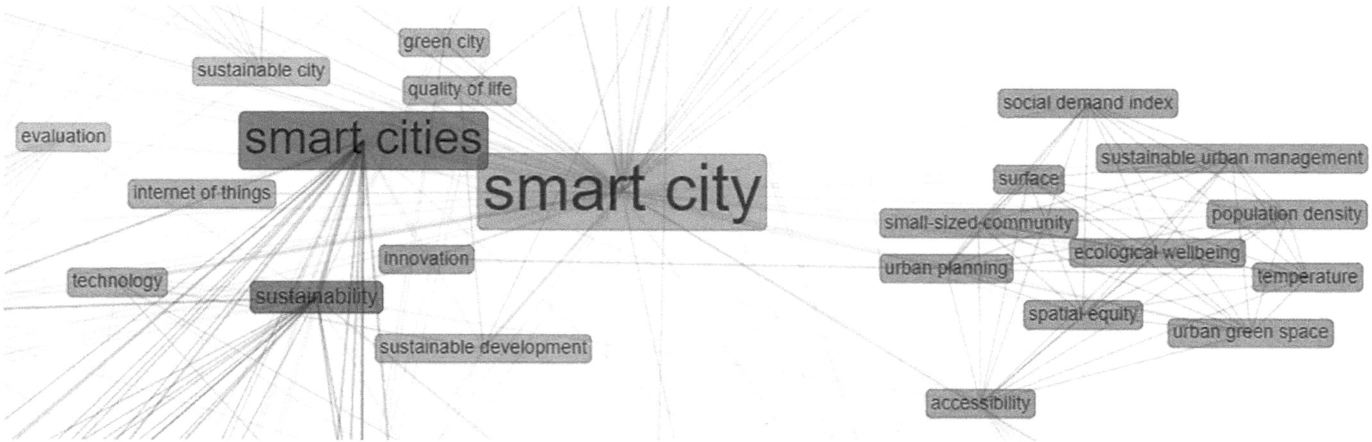

FIG. 1 Thematic network of people, sustainability, and smart city literature.

processes as a way to achieve a better quality of life, and (b) people as a target toward which to address smart interventions and their related indicators, goals, and actions. Regarding quality of life, the contribution of people to smart governance may favor a better understanding of their needs, stimulate creative contributions, and validate new solutions. The anthropic pressure is operationalized via population density and (small-sized) communities and affects indicators such as social demand, urban green spaces, spatial equity, and surface temperature useful in describing goals such as ecological well-being and accessibility. These goals may be attained through actions such as urban planning and urban management. The elements mentioned will be partnered by evidence from the literature.

Firstly, the qualitative side considers the support that smart governance offers to improve quality of life through some elements, namely, Transparency, Collaboration, Participation and Partnership, Communication and Accountability, and the relationships among them (De Guimarães et al., 2020). These elements have been proposed in Brazil to be further implemented in local policies.

Secondly, population density and the limited dimension of a community favor the achievement of ecological well-being and accessibility, with the latter referred to as an opportunity to benefit from green areas (Rahman & Zhang, 2018). Indeed, the investigation proposes four key indicators to measure the previous goals, namely, social demand index and urban green spaces as, respectively, what people ask for and what people get with reference to ecological well-being. Similarly, Bharadi and Chinagi (2019) measured the positive consequences for citizens based on population density and applied this idea to four demonstrative contexts. Moreover, accessibility is a goal affecting urban planning, to have a city suitable for initiatives for a community (Mora, Deakin, & Reid, 2018); in any event, this link was tested only with reference to sports.

3.2 People: Practical evidence

The results emerging from the theoretical analysis are supported by some practical evidence; as anticipated in the previous section, to ensure the significance and reliability of data, we chose to take into account the highest-ranked cities in the IESE Report on Cities in Motion. Regarding the relationship among people, sustainability, and the smart city, London, on top of the ranking, implemented several initiatives to improve citizens' quality of life, paying particular attention to education and cultural heritage. A similar performance was detected in Zurich, declared one of the cities with the best quality of life in the world in 2018 (Mercer Quality of Living ranking) and the second most sustainable in 2017 (Sustainable Cities Index); moreover, Zurich has one of the world's highest happiness indexes. Also, the Swiss city is one of the top cities on the Environmental Performance Index (EPI), with a very low level of pollution and CO_2 emissions, offering a healthy environment and pursuing sustainable development. In the last years, Oslo, too, has accelerated the path toward being a smart city, due to local administration meant to make it the smartest, greenest, most inclusive, and most creative city and to improve citizens' quality of life. To achieve this goal, initiatives such as zero-emission vehicles and construction sites, as well as waste management systems, have been implemented; furthermore, most services oriented toward the residents have been digitized.

3.3 Technology: Theoretical evidence

The second dataset groups the contributions on technology, sustainability, and smart cities. The results are categorized into two parts, namely, (a) technology as a mediator toward general goals, and (b) its role in achieving the implementation of more technical elements. These parts mirror most of the wide range of results from the bibliometric analysis, as depicted in the thematic network in Fig. 2.

Firstly, the ties among these contributions are based on technologies as mediating the achievement of some goals, and the thematic network connects smart cities as a topic to other themes such as the environment, urban planning, and infrastructure. Thus technology is seen as (a) a way to achieve better conditions as regards the environment, (b) an instrument to support urban planning toward smart goals, and (c) a collector of data, a tool to manage and monitor services, and therefore an infrastructure for services. In detail, scholars offered insights into these interactions, and we mention some key advances to complement this first kind of result; for instance, Turgel, Bozhko, Ulyanova, and Khabdullin (2019) stressed the advantages of smart technologies in favoring a beneficial process for the environment and also identified areas for implementing this solution in both Russia and Kazakhstan. An approach toward future implementation was used by El-Kholei & Yassine, 2019, who stressed the great support that

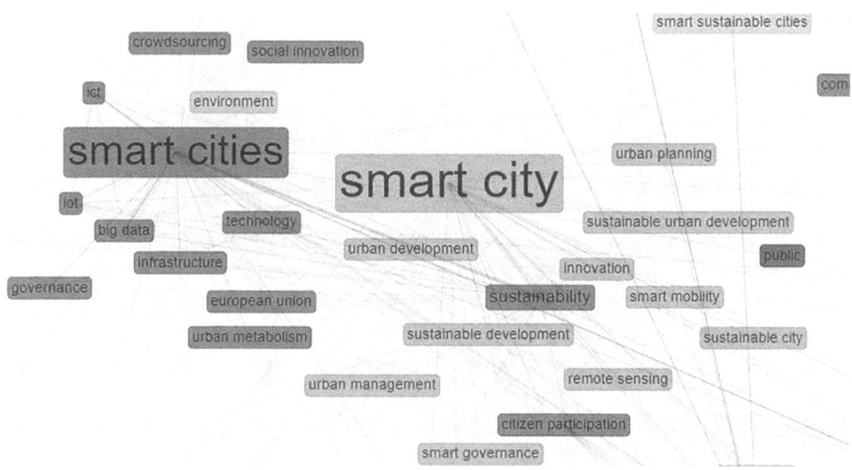

FIG. 2 Thematic network on technology, sustainability, and smart city literature.

technologies may offer to the planning process of sustainable urban contexts and also how actors' illiteracy may hinder the implementation of technology. Additionally, infrastructures play a key role in combining the resources and efforts of multiple actors, but technologies may transform these infrastructures in smarter contexts according to Selim, Yousef, and Hagag (2018). The authors discussed the contribution of technologies to projects led by a multitude of actors, as is the case in public-private partnerships for smart cities.

Secondly, the analysis led to the consideration of technology as a means of support for intervention based on more technical issues, as in the cases of ICTs, the Internet of Things, and Big Data; these three terms recall technology as an instrument to support urban governance, the achievement of sustainability, and urban metabolism. The elements are intertwined in cause-effect relationships based on what authors proposed; indeed, Webster and Leleux (2018) focused on citizens, as they activate change through ICTs in supporting the decision-making process of smart city governance. Therefore they proposed a smart governance named "technologically mediated municipal reciprocity" to stress mutual trust, citizens' engagement, and new opportunities in transforming cities. Similarly, Villegas-Ch, Palacios-Pacheco, and Luján-Mora (2019) described the support of the Internet of Things, as it may favor data collection and management, leading to the identification of efficient solutions for the sustainable development of smart cities. Shahrokni, Årman, Lazarevic, Nilsson, and Brandt (2015) took a step forward, as they moved from a proposal to an empirical observation; indeed, they used big data as key performance indicators in the assessment of sustainable interventions in Stockholm.

3.4 Technology: Practical evidence

Toronto and Amsterdam represent examples supporting the first category of results. In detail, Toronto has a very well-developed infrastructure, and the local administration paid special attention to providing adequate sanitation facilities for the whole population. Amsterdam made important investments in energy efficiency and advanced systems of automated services for public transportation such as shared bicycles. In addition, local administration settled initiatives to become Europe's first zero-emissions city by 2025 (Carbon Neutral Cities, 2019). Similarly, through the project "The City of Light," Paris promotes clean transport through the adoption of electric vehicles and bicycles and through an open innovation approach and the use of digital tools to give citizens control and access to the city's data flows. In particular, thanks to the application of the Internet of Things (IoT), Paris pursues the optimization of people and vehicle flows. Singapore is even more representative of this approach, as most of the services provided to citizens are based on technology. For example, the city provides the fastest Internet speed to its residents, although the number of digital tools used by people is very high (i.e., three mobile phones for every two inhabitants). Moreover, the city has a large number of wireless access points.

Likewise, Hong Kong faces urban challenges combining innovation and technology to enhance the effectiveness of city management and improve people's quality of life.

3.5 Policy: Theoretical evidence

The third dataset embeds journal articles dealing with policy, sustainability, and smart cities. The results are categorized in three parts, namely, (a) policy as a general topic, referring to the objects of the policy, (b) policymaking as the process leading to set policies and therefore related to interventions for development, instruments, and contexts, and (c) urban policy as a set of measures, recalling some of the previously mentioned themes, viz. planning and development. These three parts offer an overview of the most frequent occurrences from the bibliometric analysis, as can be seen in Fig. 3.

Firstly, the contributions deal with policy in general and the related objects, such as mobility, transport, planning, and API management. The debate over mobility focuses on how to achieve sustainable conditions through interventions in mobility and transport, such as the public transport system, logistics infrastructures, and accessibility to urban zones. In the same vein, planning is recalled as the process aimed at leading to change in order to achieve sustainable development of urban contexts. Additionally, API management is one of the contributions investigated, as it is a necessary process for managing data and apps as well as analyzing data and supporting the decision-making in further interventions. A scan of the literature revealed that some contributions clarifying the results mentioned earlier may expand the understanding of the linkages arising from the bibliometric analysis. As regards mobility and transport, dynamicity emerges as a key feature of a policy with reference to pricing (Jioudi, Sabir, Moutaouakkil, & Medromi, 2019); therefore the authors proposed a suitable management of spaces, movements, and means of transportation to be achieved in order to favor an optimized parking process, combining price, proximity, and

FIG. 3 Thematic network of policy, sustainability, and smart city literature.

traffic. Moreover, sustainability is addressed by policies that encourage the use of green solutions (Behrendt, 2019), as proposed in several documents issued by supranational institutions such as the European Commission. A similar approach based on documents and guidelines is adopted by Damurski and Oleksy (2018), who analyzed various methods to deduce eight basic principles—and their related importance—to be used in planning. Finally, API management is mentioned by Anthony Jr., Abbas Petersen, Ahlers, and Krogstie (2020) in infusing design science into urban studies to obtain data on mobility and derive a data-driven mobility policy. The authors tested this idea in a limited context, i.e., an infrastructure company, and proposed its implementation in urban contexts.

Secondly, policymaking is a process recalling interventions, instruments, and contexts. The most frequently recalled interventions are those related to energy, governance, and urban development; therefore policymakers and actors contributing to this task focus on actions to reduce the negative effects of energy provision and usage, as well as to define new ways to have multiple actors contributing to governance and its process. Additionally, a quite expected result is governance, as the policymakers act in tight relation with the actors dealing with governance tasks, as these two processes should be regarded as one, since they are based on common data and have mutual influence. This influence is mirrored in the choices for urban development, namely, the third interventions that emerged from the analysis. Data—as mentioned earlier—are useful to introduce instruments, too, as policymaking benefits artificial intelligence and the smart grid as new tools to support the description of the current conditions, the identification of goals, and the setting of solutions considering both previous elements. Moreover, policymaking performs its effects on limited zones, as regions, cities, or towns can be, but most of the linkages that emerged from the analysis refer to sustainable cities and smart villages. Analysis of the contributions from this dataset is useful in grasping some more meaning about policymaking; indeed, energy is another key theme in the smart transformation, leading to the need for analytical tools to support decision-making. In this vein, the authors (Pardo-García et al., 2019) proposed the potential benefits of stakeholders' participation in supporting the use of a specified platform. The inseparability of policymaking and governance is confirmed by Castelnovo (2016), who analyzed how citizens' participation describes the governance mechanisms and affects policymaking through actions described as codesign and coproduction. Participation is pivotal in policymaking even with reference to the choices for urban development, since participation guides policymaking, and scholars (Stratigea, Papadopoulou, & Panagiotopoulou, 2015) stressed the importance of starting from the city and not from the notion of smart. One more element catalyzing cities' transformation is artificial intelligence, as it can support smart service design through pilot systems and prototypes (Chui, Lytras, & Visvizi, 2018). Similar support is offered in designing smart grids as the result of a multidisciplinary approach; solutions were planned to be implemented in European cities, according to the framework that the authors took into account (Andrisano et al., 2018). With reference to contexts, two of them are often referred to in contributions, namely, sustainable cities, as they have innate features such as waste management, in contrast to environmental degradation, and eco-innovation, which can be achieved if policymaking considers proper instruments (Lagman-Bautista, 2020). Wider results can be achieved through instrumental ones, namely, smaller contexts such as smart villages, which may represent the starting point of a smart transformation, according to Fennell et al. (2018).

Thirdly, urban policy is recalled by scholars, but their contributions are aligned with the themes mentioned above, such as urban planning and urban development. For instance, Coccossis, Delladetsimas, and Niavis (2017) described participation in urban planning as crucial to unleashing the creativity necessary for successful urban policies. The role of communities and other actors in supporting policies is also described by Yigitcanlar et al. (2019) with reference to urban development based on a future considering human needs and technological opportunities.

3.6 Policy: Practical evidence

In line with the first set of results, namely, the ties between policies and their related objects, the London administration is particularly focused on sustainable and efficient transport. In general, the British capital implemented several innovative processes to improve the transport sector. For example, the "Heathrow pods" were installed, namely, capsules that work as a means of transit to connect the different terminals of Heathrow Airport. As already mentioned, Paris developed "The City of Light" project to promote clean and electric means of transport. Moreover, in the city's metropolitan area, four additional metro lines with 68 completely new interconnected stations have been implemented, all with a 100% automatic metro system. Efficiency in transport is strictly connected to energy efficiency, as emerged in the literature analysis, in which several scholars encouraged the use of green solutions. For example, Reykjavík is committed to the promotion of the use of renewable energy to reduce its dependence on fossil fuels, aimed at achieving zero carbon emissions by 2040. In addition, Reykjavík is highly ranked in the Environmental Performance Index (EPI). It is smaller than most of the other cities in the top 20 but has exemplary policies for urban planning. Indeed, Reykjavík set one of the most sustainable programs in the world, with an urban plan combining sustainable infrastructures, environmental-friendly interventions, and tools for consumption monitoring in real time.

As concerns city governance, Bern is one of the highest-ranked cities in this subcategory, as in the last years the Swiss city achieved good performance in the indexes of corruption perception, reserves per capita, and number of embassies. Economic policies are well developed in cities such as New York and Tokyo, too; the former maintained a high GDP throughout the years and a relevant number of publicly traded parent companies. Los Angeles, the second US city after New York, was one of the first to develop smart city processes, also teaming up with universities and tech companies to improve services and safeguard the natural environment. Finally, Singapore, whose local administrations have an innovation-based approach, employs new technologies in each policy or development strategy it sets and implements. Urban planning in Singapore went on for almost two decades, with the planning and development based mostly on making the city greener year after year.

4. Implications

This research combines insights from the literature, with results emerging from the observation of smart city projects throughout the world. The three levels and three elements of this research are intertwined in multiple ways. First, people participate in the debate over cities' transformation because they are interested in achieving a better quality of life and because of the engagement strategies that political institutions adopt (Lytras et al., 2021; Lytras & Visvizi, 2018; Mora et al., 2018). Moreover, people both support and constrain smart development, as the more they participate, the better the expected goals, though the bigger a city is, the harder it is for the goals to be achieved. Therefore quality of life as a goal should balance the needs of the multitude with the anthropic pressure they exert on an area (Bharadi & Chinagi, 2019). Additionally, quality of life at the microlevel may be achieved through participation and the support of technology (as in Visvizi & Lytras, 2019); our findings are aligned with this view and we stress how relevant it can be to combine participation with goals' sharing, indicators, and common actions. This combination is a strategy to be adopted to favor the sustainable development of smart cities. The last statement leads to a consideration of more details at the second level, i.e., the meso-level, with technology mediating between citizens and service providers, as technology is the way to offer services considering environmental issues, supporting more aware urban planning and favoring data collection to both assess the interventions and plan the upcoming future. Therefore the meso-level reflects the chance to achieve several goals in the unique frame of sustainability (Kumar, 2020). Our research added the twofold role of technology, with its support of service provision and outcome measures. Therefore the main novelty related to technology is its time-based role: It measures past results, monitors current ones, and elaborates both toward future planning (Shahrokni et al., 2015; Villegas-Ch et al., 2019). Planning, as a set of goals for both local and global contexts and as a process leading to decision-making, shapes the macro-level; a policy in a sustainable approach to smart cities should consider various objects, namely, services to be offered and actors to which address the services. Moreover, a policy derives from a process, and governance represents the essence of this decision-making process, with the involvement of as many actors as possible, as a development strategy for a city may not consider only a limited set of actors (Castelnovo, 2016; Visvizi & Lytras, 2018). Thus sustainable development may summarize the multiplicity of goals and actors' interests in creating changes in cities (as in Brorström et al., 2018), though what is new is the strict relation that emerged among issues, decisional processes, and implementation. Indeed, a policy is a set of goals, a set of decisions that define how to achieve these goals sustainably, and a series of actions to achieve sustainable and smart results.

5. Conclusion

This study offers a three-level analysis based on the most recurring topics in the literature—and their combination—together with evidence from the 20 cities ranked as top in the IESE Cities in Motion Index (2019) and in line with recent smart cities studies (Bifulco et al., 2016; Lytras et al., 2021; Visvizi & Lytras, 2018) as well as recent calls for research (e.g., Ghisetti, 2018). People, technology, and policy are combined, not only in theoretical contributions but also in empirical evidence, as people participate in the policymaking process through technology; moreover, policies are put into action for smart service provision to people through technology. Additionally, technology is meant to be useful for both data and actions, as data are collected during actions, i.e., services such as the deployment of smart transformation—and analyzed after actions take place. This leads to the consideration of one more tie between technology and policy, as cities plan their futures through the evidence offered by data measured by technology.

This research is based on the cities ranked in the top 20 of the IESE ranking. Therefore the insights we combined with theory are describing the best practices, while different results may emerge when one is considering other cities. It is worth noting that the considered cities cover a wide area of the world, also with different sizes and characteristics. In this regard, practical evidence shows diverse approaches, as Asian cities such as Singapore adopts an innovation-based and technology-oriented approach employing new technologies in most local initiatives, while Hong Kong uses innovation and technology as a lever to face urban challenges and improve people's quality of life. European cities, for example, Zurich and Oslo are more focused on smart services such as transport infrastructures or the provision of more efficient sources of energy. Finally, North American cities such as New York and Los Angeles are adopting more global initiatives, teaming up with universities and tech companies to improve citizens' quality of life and favor economic development.

Future research may complement these results with the adoption of a longitudinal view on city reports, as well as with the combination of additional methodologies, such as interviews with citizens and local administrators. Indeed, these interviews may favor a further understanding of the interplay among people, technology, and policy. Additionally, a further extension of the multilevel approach can lead to an understanding of the meta-level as an all-embedding view of the smart transformation of cities.

References

Albino, V., Berardi, U., & Dangelico, R. M. (2015). Smart cities: Definitions, dimensions, performance, and initiatives. *Journal of Urban Technology*, *22*(1), 3–21.

Andrisano, O., Bartolini, I., Bellavista, P., Boeri, A., Bononi, L., Borghetti, A., et al. (2018). The need of multidisciplinary approaches and engineering tools for the development and implementation of the smart city paradigm. *Proceedings of the IEEE*, *106*(4), 738–760.

Anthony, B., Jr., Abbas Petersen, S., Ahlers, D., & Krogstie, J. (2020). API deployment for big data management towards sustainable energy prosumption in smart cities—A layered architecture perspective. *International Journal of Sustainable Energy*, *39*(3), 263–289.

Aria, M., & Cuccurullo, C. (2017). Bibliometrix: An R-tool for comprehensive science mapping analysis. *Journal of Informetrics*, *11*(4), 959–975.

Bakıcı, T., Almirall, E., & Wareham, J. (2013). A smart city initiative: The case of Barcelona. *Journal of the Knowledge Economy*, *4*(2), 135–148.

Batagan, L. (2011). Indicators for economic and social development of future smart city. *Journal of Applied Quantitative Methods*, *6*(3), 27–34.

Behrendt, F. (2019). Cycling the smart and sustainable city: Analyzing EC policy documents on internet of things, mobility and transport, and smart cities. *Sustainability*, *11*(3), 763.

Bharadi, H. H., & Chinagi, G. S. (2019). Usage of water and methods of storage at households: A comparative analysis of demo zones in Hubballi-Dharwad city of Karnataka state. *The Journal of Indian Management & Strategy*, *24*(4), 11–17.

Bifulco, F., D'Auria, A., Amitrano, C. C., & Tregua, M. (2018). Crossing technology and sustainability in cities' development. *Sustainability Science*, *13*(5), 1287–1297.

Bifulco, F., Tregua, M., Amitrano, C. C., & D'Auria, A. (2016). ICT and sustainability in smart cities management. *International Journal of Public Sector Management*, *29*(2), 132–147.

Borda, A., & Bowen, J. P. (2019). Smart cities and digital culture: Models of innovation. In *Museums and digital culture* (pp. 523–549). Cham: Springer.

Bremholm, T. L. (2004). Challenges and opportunities for bibliometrics in the electronic environment: The case of the proceedings of the Oklahoma Academy of Science. *Science & Technology Libraries*, *25*(1–2), 87–107.

Brorström, S., Argento, D., Grossi, G., Thomasson, A., & Almqvist, R. (2018). Translating sustainable and smart city strategies into performance measurement systems. *Public Money & Management*, *38*(3), 193–202.

Carbon Neutral Cities. (2019). *Amsterdam climate neutral 2050 roadmap*. Available from: http://carbonneutralcities.org/wp-content/uploads/2019/12/Amsterdam-Climate-Neutral-2050-Roadmap_12072019-1.pdf.

Castelnovo, W. (2016). Co-production makes cities smarter: Citizens' participation in smart city initiatives. In *Co-production in the public sector* (pp. 97–117). Cham: Springer.

Chui, K. T., Lytras, M. D., & Visvizi, A. (2018). Energy sustainability in smart cities: Artificial intelligence, smart monitoring, and optimization of energy consumption. *Energies*, *11*(11), 2869.

Coccossis, H., Delladetsimas, P. M., & Niavis, S. (2017). The challenge of incorporating smart city activities in medium-size cities: The case of Greece. *International Journal of Services, Technology and Management*, *23*(5–6), 381–402.

D'Auria, A., Tregua, M., & Vallejo-Martos, M. C. (2018). Modern conceptions of cities as smart and sustainable and their commonalities. *Sustainability*, *10*(8), 2642.

Dameri, R. P. (2017). Smart city and digital city implementation: Two best practices in Europe. In *Smart city implementation* (pp. 109–154). Cham: Springer.

Damurski, Ł., & Oleksy, M. (2018). Communicative and participatory paradigm in the European territorial policies. A discourse analysis. *European Planning Studies*, *26*(7), 1471–1492.

De Guimarães, J. C. F., Severo, E. A., Júnior, L. A. F., Da Costa, W. P. L. B., & Salmoria, F. T. (2020). Governance and quality of life in smart cities: Towards sustainable development goals. *Journal of Cleaner Production, 253*, 119926.

De Jong, M., Joss, S., Schraven, D., Zhan, C., & Weijnen, M. (2015). Sustainable–smart–resilient–low carbon–eco–knowledge cities; making sense of a multitude of concepts promoting sustainable urbanization. *Journal of Cleaner Production, 109*, 25–38.

El-Kholei, A. O., & Yassine, G. (2019). The mirage of smart sustainable cities in the Arab region. *Open House International, 44*(1), 8–16.

Fennell, S., Kaur, P., Jhunjhunwala, A., Narayanan, D., Loyola, C., Bedi, J., et al. (2018). Examining linkages between smart villages and smart cities: Learning from rural youth accessing the internet in India. *Telecommunications Policy, 42*(10), 810–823.

Ghisetti, C. (2018). Smart cities, innovation and sustainability: Which role for cities in fostering "green" entrepreneurship? In *Sustainability in innovation and entrepreneurship* (pp. 31–45). Cham: Springer.

Giffinger, R., Fertner, C., Kramar, H., & Meijers, E. (2007). City-ranking of European medium-sized cities. *Cent. Reg. Sci. Vienna UT*, 1–12.

Gilderbloom, J. I. H., Squires, G. D., Riggs, W., & Čapek, S. (2017). Think globally, act locally: Neighbourhood pollution and the future of the earth. *Local Environment, 22*(7), 894–899.

Hollands, R. G. (2015). Critical interventions into the corporate smart city. *Cambridge Journal of Regions, Economy and Society, 8*(1), 61–77.

IESE. (2019). *IESE cities in motion index*. Retrieved from: https://blog.iese.edu/cities-challenges-and-management/2019/05/10/iese-cities-in-motion-index-2019/.

Ishida, T. (2017, April). Digital city, smart city and beyond. In *Proceedings of the 26th international conference on world wide web companion* (pp. 1151–1152).

Jayashree, P., Hamza, F., El Barachi, M., & Gholami, G. (2019, June). Inclusion as an enabler to sustainable innovations in smart cities: A multi-level framework. In *2019 4th International conference on smart and sustainable technologies (SpliTech)* (pp. 1–9). IEEE.

Jioudi, B., Sabir, E., Moutaouakkil, F., & Medromi, H. (2019). Congestion awareness meets zone-based pricing policies for efficient urban parking. *IEEE Access, 7*, 161510–161523.

Johnson, J. (2012). Cities: Systems of systems of systems. In *Complexity theories of cities have come of age* (pp. 153–172). Berlin, Heidelberg: Springer.

Klopp, J. M., & Petretta, D. L. (2017). The urban sustainable development goal: Indicators, complexity and the politics of measuring cities. *Cities, 63*, 92–97.

Kohlbacher, F. (2006). The use of qualitative content analysis in case study research. *Qualitative Social Research, 7*(1). http://nbn-resolving.de/urn:nbn:de:0114-fqs0601211. art. 21.

Komninos, N. (2013). What makes cities intelligent? In *Smart cities* (pp. 89–107). Abingdon, UK: Routledge.

Kumar, T. V. (2020). Smart environment for smart cities. In *Smart environment for smart cities* (pp. 1–53). Singapore: Springer.

Lagman-Bautista, J. (2020). Crafting a theoretical framework on waste management: A case for sustainable cities. *International Journal of Geomate, 18*(68), 80–86.

Lytras, M. D., & Visvizi, A. (2018). Who uses smart city services and what to make of it: Toward interdisciplinary smart cities research. *Sustainability, 10*, 1998.

Lytras, M. D., Visvizi, A., Chopdar, P. K., Sarirete, A., & Alhalabi, W. (2021). Information management in smart cities: Turning end users' views into multi-item scale development, validation, and policy-making recommendations. *International Journal of Information Management, 56*(February), 102146.

Mora, L., Deakin, M., & Reid, A. (2018). Stylized facts on smart specialisation research. In *Economic and social development: Book of proceedings* (pp. 176–185).

Mora, H., Gilart-Iglesias, V., Pérez-del Hoyo, R., & Andújar-Montoya, M. D. (2017). A comprehensive system for monitoring urban accessibility in smart cities. *Sensors, 17*(8), 1834.

Nam, T., & Pardo, T. A. (2011, June). Conceptualizing smart city with dimensions of technology, people, and institutions. In *Proceedings of the 12th annual international digital government research conference: Digital government innovation in challenging times* (pp. 282–291).

Neirotti, P., De Marco, A., Cagliano, A. C., Mangano, G., & Scorrano, F. (2014). Current trends in smart city initiatives: Some stylised facts. *Cities*, *38*, 25–36.

Nižetić, S., Djilali, N., Papadopoulos, A., & Rodrigues, J. J. (2019). Smart technologies for promotion of energy efficiency, utilization of sustainable resources and waste management. *Journal of Cleaner Production*, *231*(10), 565–591.

Pardo-García, N., Simoes, S. G., Dias, L., Sandgren, A., Suna, D., & Krook-Riekkola, A. (2019). Sustainable and resource efficient cities platform–SureCity holistic simulation and optimization for smart cities. *Journal of Cleaner Production*, *215*, 701–711.

Rahman, K. M., & Zhang, D. (2018). Analyzing the level of accessibility of public urban green spaces to different socially vulnerable groups of people. *Sustainability*, *10*(11), 3917.

Rudie, K., & Wonham, W. M. (1991, June). Think globally, act locally: Decentralized supervisory control (pp. 898–903). Boston, USA: IEEE.

Schilling, J., & Logan, J. (2008). Greening the rust belt: A green infrastructure model for right sizing America's shrinking cities. *Journal of the American Planning Association*, *74*(4), 451–466.

Selim, A. M., Yousef, P. H., & Hagag, M. R. (2018). Smart infrastructure by (PPPs) within the concept of smart cities to achieve sustainable development. *International Journal of Critical Infrastructures*, *14*(2), 182–198.

Shahrokni, H., Årman, L., Lazarevic, D., Nilsson, A., & Brandt, N. (2015). Implementing smart urban metabolism in the Stockholm Royal Seaport: Smart city SRS. *Journal of Industrial Ecology*, *19*(5), 917–929.

Soe, R. M. (2020). Mobility in smart cities: Will automated vehicles take it over? In *Smart governance for cities: Perspectives and experiences* (pp. 189–216). Cham: Springer.

Stratigea, A., Papadopoulou, C. A., & Panagiotopoulou, M. (2015). Tools and technologies for planning the development of smart cities. *Journal of Urban Technology*, *22*(2), 43–62.

Tanguay, G. A., Rajaonson, J., Lefebvre, J. F., & Lanoie, P. (2010). Measuring the sustainability of cities: An analysis of the use of local indicators. *Ecological Indicators*, *10*(2), 407–418.

Turgel, I., Bozhko, L., Ulyanova, E., & Khabdullin, A. (2019). Implementation of the smart city technology for environmental protection management of cities: The experience of Russia and Kazakhstan. *Environmental and Climate Technologies*, *23*(2), 148–165.

Villegas-Ch, W., Palacios-Pacheco, X., & Luján-Mora, S. (2019). Application of a smart city model to a traditional university campus with a big data architecture: A sustainable smart campus. *Sustainability*, *11*(10), 2857.

Visvizi, A., & Lytras, M. D. (2018). Rescaling and refocusing smart cities research: From mega cities to smart villages. *Journal of Science and Technology Policy Management*, *9*(2), 134–145.

Visvizi, A., & Lytras, M. (Eds.). (2019). *Smart cities: Issues and challenges: Mapping political, social and economic risks and threats*. Amsterdam, The Netherlands: Elsevier.

Visvizi, A., Lytras, M. D., Damiani, E., & Mathkour, H. (2018). Policy making for smart cities: Innovation and social inclusive economic growth for sustainability. *Journal of Science and Technology Policy Management*, *9*(2), 126–133.

Webster, C. W. R., & Leleux, C. (2018). Smart governance: Opportunities for technologically-mediated citizen co-production. *Information Polity*, *23*(1), 95–110.

Yigitcanlar, T., Kamruzzaman, M., Foth, M., Sabatini-Marques, J., da Costa, E., & Ioppolo, G. (2019). Can cities become smart without being sustainable? A systematic review of the literature. *Sustainable Cities and Society*, *45*, 348–365.

Yigitcanlar, T., & Lönnqvist, A. (2013). Benchmarking knowledge-based urban development performance: Results from the international comparison of Helsinki. *Cities*, *31*, 357–369.

Chapter 4

Technology and SDGs in smart cities context

Chrysaida-Aliki Papadopoulou
School of Rural and Surveying Engineering, National Technical University of Athens, Athens, Greece

1. Introduction

SDGs were adopted by the UN Member States in 2015 and constitute part of the Agenda 2030. They represent the blueprint to achieve a more sustainable, prosperous, and equitable future for all (United Nations, 2020). SDGs succeeded the Millennium Development Goals (MDGs) but were designed on a different basis. Such new rationale proposes that more sustainability is possible to be achieved by identifying all countries as "developing" that need to be further transformed (Biermann, Kanie, & Kim, 2017). Thus, a starting point for the implementation of SDGs is the exploration of the state of the art, the investigation of specific needs and problems of each country, and the design of local-oriented solutions under the general context of SDGs. Particular characteristics, weaknesses, and comparative advantages should be considered. Moreover, SDGs are strongly interrelated as the accomplishment of a specific goal depends on the progress of the rest. Nilsson, Griggs, and Visbeck (2016) and Weitz, Carlsen, Nilsson, and Skånberg (2018) reported on such interactions by using a seven-point typology and assessing synergies and possible conflicts among SDGs. The assessment revealed critical synergies and highlighted the need for a combined application of SDGs in order to attain the highest level of performance.

According to the most recent forecasts of United Nations, urban population is expected to be further increased in the future and by 2030, 6 out of 10 people will live in cities. Urban centers will sustain additional pressures and innovative solutions should be investigated through the adoption of integrated approaches and modern technologies, targeting at the efficient management of human and natural capital. SDG 11: "Make cities and human settlements inclusive, safe, resilient, and sustainable" focuses on the management of urban assets by involving and intertwining the concepts of "inclusion," "safety," "resilience," and "sustainability." Special emphasis is placed on citizens' standards of living

Smart Cities and the UN SDGs. https://doi.org/10.1016/B978-0-323-85151-0.00004-X

and the confrontation of climate change impacts through adaptation and mitigation strategies. Strategies that will make cities more sustainable and ensure socioeconomic balance are also at the forefront of urban evolvement (Visvizi & Lytras, 2019).

Smart cities represent a promising pattern of urban development as they bring the potential to promote sustainability and human well-being through the extensive use of ICT. ICT comprises the driving force of smart cities allowing for data collection and management, data integration, and acceleration of the accomplishment of future goals and prospects. The role that technology may play to the successful implementation and achievement of SDG 11 has been investigated by several researchers, in an effort to clarify the relation between smart city and sustainable urban development as well as the role of smart technologies in achieving sustainability goals (Allam & Newman, 2018; Anand & Navío-Marco, 2018; Angelidou et al., 2018; Huovila, Bosch, & Airaksinen, 2019; Yigitcanlar et al., 2019). However, such issues are rather recent and a plethora of research articles, focusing on the progress of SDG 11 and the performance of relevant target-specific indicators in different urban contexts, are expected to be available in the near future.

This chapter focuses on the role that technology may play toward the achievement of SDG 11 in smart urban environments. More specifically, SDG 11, relevant targets, and corresponding indicators are analytically presented. Then, technologies supporting the development of a smart city and shaping its backbone are delineated. The contribution of smart technologies to the accomplishment of SDG 11 and each specific objective is also explored. Finally, a conclusive discussion takes place.

2. SDG 11: Concepts, objectives, and aspirations

SDG 11 "Make cities and human settlements inclusive, safe, resilient and sustainable" builds on the concepts of urban sustainability, citizens' well-being, social coherence, participation, and efficient use of available resources. Emphasis is placed on the limitation of poverty by reducing urban population living in slums; the decrease of GHG emissions and improvement of air quality; the accessibility to public transportation means and open public/green spaces; the efficient waste management; the allocation and adequacy of residential areas and infrastructures; the limitation of urban sprawl and consequent environmental impacts; the creation of job opportunities; the elimination of vulnerability to natural disasters; the establishment of modern and sustainable buildings; and the reinforcement of telecommunications sector. In this context, the main pressures that need to be effectively addressed, concern the sectors of freshwater supply, sewage, and public health; the sector of energy; the infrastructures that are susceptible to natural disasters; the protection of open and green spaces; the regulation of land uses; and the sector of urban mobility.

Under this framework, the importance of sustainable urban planning is highlighted (Almeida, Smart, & Davey, 2018) and the potential of spatial planning in supporting the design of robust decisions and decisive interventions is also underlined. National/regional policies and local-oriented development plans constitute the key tools for designing interventions and targeted measures that will further support the successful implementation of SDG 11 as they introduce regulatory mechanisms, setting the terms and conditions under which several economic sectors may be evolved.

SDG 11 represents a general framework promoting sustainable urban development. It is further specialized into a number of targets, reflecting sectoral priorities and taking into account issues related to the management of pressures and problems of contemporary urban areas. Each target is accompanied by a relevant indicator measuring its performance and accomplishment level.

More analytically, Target 11.1 focuses on the sector of housing and the reduction of urban population living in slums. It aims at improving standards of living through the provision of adequate housing infrastructures. Target 11.2 places emphasis on providing access to public transportation services through the establishment of sustainable and safe transportation systems, accessible by all and especially by the sensitive social groups (e.g., people with disabilities). Target 11.3 highlights issues related to social inclusion and participatory actions when spatial decision-making, urban planning and land use management initiatives take place. Target 11.4 considers the need to protect natural and cultural heritage on national, regional, and local scales through the establishment of funding schemes that will support the implementation of relevant investments. Target 11.5 concerns the elimination of disasters and the implicit economic losses, the reduction of deaths and people affected by natural disasters as well as the protection and upgrading of vulnerable infrastructures. Target 11.6 pays attention to the improvement of air quality, the sustainable management of waste, and the limitation of consequent environmental impacts. Target 11.7 prioritizes the sustainable management of the built environment and the need for providing access to green public spaces to all, especially to sensitive social groups. Target 11A builds on the efficient management of social, economic, and environmental resources through the constitution of national and regional plans that will support the integrated development of urban, suburban, and rural regions. Target 11B focuses on the formulation of risk reduction strategies at both national and local levels, considering climate change impacts and the respective need for adaptation and mitigation policies. Finally, Target 11C refers to the establishment of resource-efficient buildings in the least developed countries by offering the necessary technical and financial support.

Indicative indicators, measuring the level of achievement of the abovementioned targets, include (United Nations, 2020): the proportion of urban population living in slums; the amount of population having access to public transport; the ratio of land consumption rate to population growth rate; the proportion of cities with direct participatory and democratic initiatives as to urban planning and management; the total expenditure per capita spent for the protection of cultural and natural heritage; the number of people affected by disasters; the direct disaster economic losses; the proportion of solid waste regularly managed by cities; the average share of the built-up area that is open for public use; the proportion of financial support to the least developed countries, etc.

At this point, it should be mentioned that the implementation of SDG 11 and its relevant targets presupposes their adaptation and "translation" into local contexts, needs, and peculiarities. Thus, further specialization is required when it comes to local conditions where particular characteristics, weaknesses, and comparative advantages should be considered. The first level of specialization is the national one, where all SDGs, targets, and indicators should be adapted and contextualized (Arslan, Durak, & Aytac, 2017; Koch & Krellenberg, 2018). A second specialization level is that of regional scale (e.g., NUTS 2 level) while a third and more local-oriented one should take into account the particular features of local communities (e.g., municipalities).

The successful implementation of SDG 11 and relevant targets will be further strengthened through the creation of collaborative partnerships, working synergistically toward the attainment of common ends. Coordinated initiatives constitute a significant dimension of SDG 11 as they contribute to raising awareness, exploring broadly accepted solutions, knowledge sharing, and dissemination. Under a participatory planning context, multi-stakeholder partnerships and voluntary commitments may efficiently address issues having to do with access to basic services and transportation services, housing allocation, sustainable waste management, green-facility management, rational use of water resources, assessment of future risks, and proactive planning. Consequently, local partnerships represent an essential element toward the accomplishment of SDG 11 and a concept that interrelates SDG 11 with SDG 17 "Partnerships for the Goals" (MacDonald, Clarke, Huang, Roseland, & May Seitanidi, 2018).

In a smart city context, the extensive use of modern technologies and the familiarization of citizens with such tools may further support the implementation of plans targeting at the achievement of SDG 11. Data collection and data management tools, Geographical Information Systems (GIS), tools supporting online participation, living labs, and Decision Support Systems (DSS) are among the main technologies used in smart urban environments in order to support the smart transition of cities and citizens' well-being. Initiatives toward the achievement of SDG 11 could benefit from such technologies facilitating the implementation of relevant interventions included in the priorities of the goal.

3. Technologies supporting the smart dimension of cities

Smart cities constitute a modern and popular trend of urban development that promotes the sustainable management of natural and human resources through the exploitation of ICT and technological advancements, cutting horizontally the functions of a city (Papadopoulou & Hatzichristos, 2020). The fundamental axes upon which the evolvement of a smart city is based on, simultaneously "shaping" its backbone, include: smart governance, smart living, smart mobility, smart people, smart environment, and smart economy (Giffinger et al., 2007). The term "smart" implies the adoption of technologies supporting the efficient use of available economic, social, and natural capital in order to address existing and forecasted problems. It entails the modernization of the city's ICT infrastructures and the extensive use of technological tools in combination with traditional services and infrastructures in order citizens' standards of living to be improved under the framework of sustainable policy initiatives.

Although a vast number of definitions reflecting on the particular attributes of a smart city have been proposed in the literature, a common and unified definition, corresponding to all kinds of cities, has not been generated (Papadopoulou & Hatzichristos, 2020; Visvizi & Lytras, 2018). This is because of the diverse characteristics forming the profile of each city, the variety of cities' size and population density as well as the differences concerning the cultural background and access to resources. However, most definitions involve concepts such as "human asset," "traditional infrastructures," "available resources," "ICT," "citizens' well-being," "social inclusion," "urban challenges" (Batty et al., 2012; Caragliu, Del Bo, & Nijkamp, 2009; Dameri, 2013; Glasmeier & Christopherson, 2015; Lombardi, 2011; Manville et al., 2014; Mitchell, Villa, Stewart-Weeks, & Lange, 2013; Ramaprasad, Sánchez-Ortiz, & Syn, 2017) in an effort to formulate a concrete, integrated, and representative definition of the smart city. Under this framework, traditional infrastructure and ICT are innovatively and intelligently combined under a "smart" rationale targeting at improving cities' economic prosperity, social cohesion, resource accessibility, and environmental protection.

Technologies setting the ground for the smart transition of contemporary cities support a huge spectrum of processes related to data collection and management, urban planning, participatory initiatives, and urban decision-making. Among the most popular ones are GIS/web-GIS, Participatory GIS/ Public Participatory GIS (PGIS/PPGIS), living labs and crowdsourcing platforms, cloud computing tools, Location-Based Services (LBS), Augmented Reality (AR) applications, Internet of things/Internet of everything (IoT/IoE), DSS/Spatial DSS (SDSS) and sensors, enabling the design and operation of smart infrastructures and facilities (e.g., smart energy grids, smart waste management applications, smart parking facilities, smart houses, etc.). Such technologies support data collection and storage (e.g., sensors, cloud computing, crowdsourcing platforms), data management (e.g., GIS, AR, LBS), innovation development (e.g., living labs), visualization/geo-visualization of spatial data and spatial relations (e.g.,GIS), stakeholder engagement in decision-making (e.g., PGS/PPGIS), integrated assessment of future scenarios (DSS/SDSS), integration of different technological tools (IoT), etc.

More specifically, GIS allow for the visualization of spatial data and spatial relations and the implementation of spatial analysis processes in order to produce "new" and innovative knowledge. They also support spatial decision-making through the production of decision maps, enabling the visual interpretation of relevant outcomes. Apart from the aforementioned potentialities, GIS/web-GIS, and PGIS/PPGIS enhance "visual thinking" and comprehension of geographic reality, facilitate the so-called "argumentation mapping," enable the dissemination of geographic knowledge, and increase mutual understanding. Moreover, they strengthen online participatory decision-making as users have the opportunity to interact with relevant applications, post comments, and annotations, make suggestions, and propose possible alternative solutions to existing problems based on their specific knowledge, preferences, and expertise. The adoption of such visualization tools reinforces visual communication skills and perception of a study problem.

LBS exploit geographical data and information in order to offer users services including a spatial reference. They integrate locations in the geographical space with various city services (e.g., health services, entertainment facilities, tourist services, etc.). Furthermore, LBS support emergency management by processing information emanating from citizens' mobiles (Aloudat, Michael, Chen, & Al-Debei, 2014), traffic management and urban planning (Huang, Gartner, Krisp, Raubal, & Van de Weghe, 2018), and access to all kind of geo-referenced urban facilities. As for AR, it is based on 3D representations of objects embedded in 3D representations of reality and enables interaction with real world through the use of computers. Users can explore an area of interest and experiment with different options for future development.

Living labs and crowdsourcing mobilize collective thinking and collaborative innovation production. Crowdsourcing builds on the "wisdom of the crowd" as it concerns the "outsourcing" of existing problems through the use of interactive platforms and the collection of candidate solutions offered by the "crowd." It thus exploits "Masscapital" and a digital community is established, offering its expertise toward seeking possible solutions to problems published online (Garrigos-Simon & Narangajavana, 2015; Gatautis & Vitkauskaite, 2014; Thuan, Antunes, & Johnstone, 2016; Whitla, 2009). Living labs constitute open innovation laboratories promoting synergetic and user-driven innovation production as well as knowledge/expertise dissemination for addressing problems related to several economic sectors evolving in a city (Eriksson, Niitamo, & Kulkki, 2005; Letaifa, 2015; Papadopoulou & Giaoutzi, 2017; Schaffers et al., 2011; Voytenko, McCormick, Evans, & Schliwa, 2016).

Cloud computing supports the storage of vast amounts of data and information, the management of peak loads, and the provision of services to a broad and distributed network of computers. It allows the delivery of applications as services via the Internet (Armbrust et al., 2010) and resource management on the fly (Ali, Khan, & Vasilakos, 2015). It is usually combined with IoT in order to boost the development of improved applications. IoT concerns the intelligent communication of devices and the connection of "things" (Li, Xu, & Zhao, 2015); devices turn out to be smarter and the available information is enriched (Ray, 2018). IoE goes a step beyond by suggesting a broader connection pattern, integrating devices, people, processes, and data (Hussain & Hussain, 2015), and enabling the share of content (Di Martino, Li, Yang, & Esposito, 2018) among all connected entities.

Regarding DSS/SDSS, they offer the chance to integrate knowledge derived from multiple sources into a single knowledge base, make assumptions as to possible alternative solutions for a study problem, assess relevant impacts, and formulate informative decisions. Moreover, they promote the participation of several interested parties and collaborative decision-making.

Finally, a significant "structural element" of smart cities is sensors. Sensors offer the potential to collect any type of data, transmit data (data sharing), and monitor city functions. Sensor data support the sustainable management of resources and critical infrastructures, and set the ground for designing more effective urban plans.

The abovementioned technologies are extensively utilized in contemporary cities and "shape" their smart profile. A significant number of applications, facilitating citizens' daily life, have been developed by exploiting their potentialities while cities may further benefit from technological advancements in order to attain future goals and aspirations. Low-carbon transitions, climate change adaptation and mitigation challenges, housing development, waste management, social inclusion, management of disaster risks, and upgrading of transportation networks constitute indicative sectors that can be fully assisted by technologies supporting data collection, data management, data processing, and generation of innovative knowledge. In this context, modern technologies may offer an additive value to the accomplishment of SDG 11 and relevant targets in smart urban environments where citizens' well-being, economic development, and environmental protection are intelligently addressed.

4. Technological innovations serving the accomplishment of SDG 11

SDG 11 seeks to enhance urban prosperity by promoting social inclusion, resilience, and environmental sustainability. It establishes a human-centered perspective and encourages participatory initiatives toward the accomplishment of common ends. Future challenges are expected to be addressed through the adoption of holistic approaches engaging society, resource use, economic assets, and climate change adaptation under a common framework of sustainable urban development. The priorities of SDG 11 and its specific targets refer to the affordability of housing sector, the accessibility to public transport, the extensive citizen participation in urban planning processes, the protection of cultural heritage, the limitation of natural disasters and consequent economic losses, the reduction of GHGs, the management of urban/municipal waste, the accessibility to green public spaces, the integrated spatial development, the efficient use of resources under climate change conditions, and the establishment of resilient buildings in the least developed countries.

Smart cities constitute the ideal pattern of urban development that may boost the implementation and successful accomplishment of SDG 11 priorities. This is due to the available ICT infrastructure bringing the potential to facilitate the design and application of urban plans which aim at the confrontation of future challenges. Among the most critical factors contributing to the constitution of effective city plans are data availability and data management. Especially, when it comes to the estimation of targets' performance (indicators), data availability is crucial (Klopp & Petretta, 2017) in order to precisely assess the progress of implementation. Sensors, LBS, and crowdsourcing platforms support the collection of any type of data such as traffic data, waste disposal data, climatic data, data related to the level of GHG emissions, data concerning the location of several facilities in the city as well as emergency data (e.g.,damages in water/wastewater networks, traffic accidents, etc.). Such data may be used by GIS, AR platforms, and DSS/SDSS in order to investigate existing problems, explore possible future solutions, and assess their impacts. Decision maps put an additive value to the whole venture as they allow for the visualization, argumentation mapping, and analysis of all involved parameters. Issues like allocation of residential areas, bus stops, green spaces, industrial activities, educational institutions, etc. may be effectively managed through the extensive adoption of such technologies supporting the undertaking of informative decisions. IoT and IoE set the basis for the broad communication and networking of data collection/data management devices, urban procedures, and citizens. They shape the ubiquitous architecture of smart cities where a connected network of "everything" offers the opportunity to tune actions and efforts toward a sustainable and fair future urban development.

More analytically, the sector of housing development may be fully supported by geospatial technologies, enabling spatial analysis, and exploration of the most suitable areas for residential use. Land use maps allow for gaining a deep insight into the study area, understanding spatial patterns, and designing solutions for upgrading deprived areas (e.g., slums). Moreover, 4G and 5G networks facilitate information sharing and communication while they also create job opportunities through teleworking. Telematics, GNSS-based web applications, LBS, smart parking services, etc. support safety and sustainability of transportation services and improve citizens' accessibility potential. PGIS/PPGIS, living labs, crowdsourcing, and any kind of participatory platforms offer citizens the chance to get involved in several decision-making processes affecting their daily life. Citizens co-shape/codesign the future evolvement of "their" city/neighborhood while they also indulge into adopted technologies and take an active role during innovation production. Initiatives concerning knowledge dissemination as to cultural heritage protection may also be promoted through the exploitation of ICT such as virtual museums, virtual guides to ancient treasures, etc. Regarding natural disasters, the extensive use of sensors and information sharing systems contribute to a better tuning of emergency services and provide forecasts that enhance proactive planning. Smart energy grids and smart transportation services support the transition to low-carbon economies. In the same line smart waste management solutions, usually supported by cloud-based software, sensors, and IoT, allow monitoring of waste disposal, on-time waste collection and assessment of bins' fill level. Finally, smart technologies contribute to the modernization of the building sector by providing energy-efficient solutions, water-saving equipment, and smart appliances that reinforce the effective and sustainable use of resources.

Indicative examples originate from: Mora, Gilart-Iglesias, Pérez-del Hoyo, Andújar-Montoya, and Compañ-Gabucio (2016), concerning the design of a system that analyzes the level of citizens' accessibility to public transport by combining GPS, GIS, smart sensing, and cloud computing technologies; Riganti (2017), focusing on the potential of virtual environments, virtual globes, and crowdsourcing for the preservation of cultural heritage; Mora, Gilart-Iglesias, Pérez-del Hoyo, and Andújar-Montoya (2017), introducing a system that deals with the requirements of people with movement disabilities by exploiting the benefits of sensors and cloud computing; Plageras, Psannis, Stergiou, Wang, and Gupta (2018), referring to the establishment of green smart buildings through the combined use of IoT, cloud computing, big data, and sensors; Dircke and Molenaar (2010), reporting on the development of a serious flood management game for Rotterdam, targeting at water management and climate change adaptation; the Amsterdam Institute for Advanced Metropolitan Solutions (2020), concerning the evolvement of joint stakeholder efforts in order urban challenges to be addressed through collaborative innovation production; Shyam, Manvi, and Bharti (2017), dealing with smart waste management by providing intelligence to waste bins using IoT and sensors.

5. Discussion and conclusions

Contemporary cities are facing critical challenges that need to be effectively addressed in order to meet future sustainability targets. In the forthcoming decades, the urban population and urban sprawl are expected to be further increased. SDG 11 sets the main priorities that "cities of tomorrow" should fulfill by placing emphasis on citizens' standards of living, rational use of natural resources, sophisticated management of economic assets, and inclusion of social capital in urban planning processes. Smart cities constitute a recent and promising paradigm of urban development that brings the potential to strongly support the accomplishment of SDG 11 through the extensive and ubiquitous use of ICT.

Literature review shed light on the most popular technologies, serving a broad spectrum of city functions such as traffic administration, energy provision, smart water management, smart building, and public participation. Representative applications have also been elicited, indicating that smart initiatives may set the basis toward the achievement of SDG 11.

Smart cities' priorities and SDG 11-related targets include common aspirations for the future as they build on the concept of sustainability and its main pillars namely, environmental protection, economic development, and social cohesion. Thus, smart urban environments set the ground for the specialization of SDG 11 into local-specific contexts and offer the technological background supporting the intensification of efforts toward its accomplishment. In addition, they represent ideal environments for evaluating the progress of relevant indicators as smart technologies support the collection, storage, and analysis of essential data as well as the production of "new" and innovative knowledge. Such technologies bring the potential to create a measurement and monitoring framework in order SDG 11 indicators to be calculated and the advancement of SDG 11 to be assessed. Thus, performance assessment is enhanced and decision-making processes regarding the institution of more efficient policies toward the accomplishment of inclusivity, urban sustainability, and resilience are strengthened. A technology-driven pattern of urban development boosts the design of smart solutions and the adoption of effective monitoring methods through the specialization of targets and indicators into local contexts, the use of multiple data sources, the production of standardized and comparable data, the integration of data referring to several sectors, and the establishment of open data platforms. Under such a rationale, the United Nations recognize that the evolvement of future sustainable cities requires the development of high-standard ICT infrastructures that will enrich the available data and "guarantee" for their quality. Modern ICT and the extensive use of smart applications may strongly contribute to face existing problems such as lack of precise data and information that are necessary for monitoring the progress of SDG 11 and for reviewing current ineffective policies and already implemented practices. The availability and systematic analysis of data concerning poverty, criminality, access to public transport, land use management, expenditures per capita, economic losses, natural/economic impacts of disasters, solid waste management, sustainable urban planning, and financial support to the least developed countries will allow for a more comprehensive assessment of SDG 11 achievement level, shed light on possible shortcomings, reveal existing conflicts and trade-offs, and set the basis for the design of robust solutions and informative policy decisions that will accelerate the accomplishment of SDG 11 and its specific targets.

Consequently, smart initiatives may boost the upgrading of deprived suburbs, support proactive planning and reduction of possible risks, increase awareness, enhance participatory actions, promote the establishment of resilient buildings, upgrade transportation services, and strengthen the implementation of climate change adaptation plans and mitigation strategies. They also support data acquisition and facilitate the assessment of indicators as to the advancement of SDG 11, e.g., proportion of urban population living in slums, amount of population having access to public transport, total expenditure per capita spent for the protection of cultural and natural heritage, direct disaster economic losses, proportion of solid waste regularly managed by cities, etc. However, a human-centric approach should form the background of all such efforts as people "shape" the identity of a city, and smart cities cannot be evolved without smart people and active citizenship.

References

Ali, M., Khan, S. U., & Vasilakos, A. V. (2015). Security in cloud computing: Opportunities and challenges. *Information Sciences*, *305*, 357–383. https://doi.org/10.1016/j.ins.2015.01.025.

Allam, Z., & Newman, P. (2018). Redefining the smart city: Culture, metabolism and governance. *Smart Cities*, *1*(1), 4–25. https://doi.org/10.3390/smartcities1010002.

Almeida, A. C. L., Smart, J. C. R., & Davey, P. (2018). Can learned experiences accelerate the implementation of sustainable development goal 11? A framework to evaluate the contributions of local sustainable initiatives to delivery SDG 11 in Brazilian municipalities. *European Journal of Sustainable Development*, *7*(4), 517–530. https://doi.org/10.14207/ejsd.2018.v7n4p517.

Aloudat, A., Michael, K., Chen, X., & Al-Debei, M. M. (2014). Social acceptance of location-based mobile government services for emergency management. *Telematics and Informatics*, *31*(1), 153–171. https://doi.org/10.1016/j.tele.2013.02.002.

Amsterdam Institute for Advanced Metropolitan Solutions. (2020). Retrieved from: https://www.ams-institute.org/how-we-work/living-labs/.

Anand, P. B., & Navío-Marco, J. (2018). Governance and economics of smart cities: Opportunities and challenges. *Telecommunications Policy*, *42*(10), 795–799. https://doi.org/10.1016/j.telpol.2018.10.001.

Angelidou, M., Psaltoglou, A., Komninos, N., Kakderi, C., Tsarchopoulos, P., & Panori, A. (2018). Enhancing sustainable urban development through smart city applications. *Journal of Science and Technology Policy Management*, *9*(2), 146–169. https://doi.org/10.1108/JSTPM-05-2017-0016.

Armbrust, M., Fox, A., Griffith, R., Joseph, A. D., Katz, R., Konwinski, A., et al. (2010). A view of cloud computing. *Communications of the ACM*, *53*(4), 50–58.

Arslan, T. V., Durak, S., & Aytac, D. O. (2017). Attaining SDG 11: Can sustainability assessment tools be used for improved transformation of neighbourhoods in historic city centers? *Natural Resources Forum*, *40*(4), 180–202. https://doi.org/10.1111/1477-8947.12115.

Batty, M., Axhausen, K., Fosca, G., Pozdnoukjov, A., Bazzani, A., Wachowicz, M., et al. (2012, October). *Smart cities of the future (paper no. 188)*. London, United Kingdom: University College London (UCL), Centre for Advanced Spatial Analysis (CASA).

Biermann, F., Kanie, N., & Kim, R. (2017). Global governance by goal-setting: The novel approach of the UN sustainable development goals. *Current Opinion in Environmental Sustainability*, *26–27*, 26–31. https://doi.org/10.1016/j.cosust.2017.01.010.

Caragliu, A., Del Bo, C., & Nijkamp, P. (2009). *Smart cities in Europe (series research memoranda 0048)*. Amsterdam, Netherlands: VU University Amsterdam, Faculty of Economics, Business Administration and Econometrics.

Dameri, R. P. (2013). Searching for smart city definition: A comprehensive proposal. *International Journal of Computers and Technology*, *11*(5), 2544–2552. https://doi.org/10.24297/ijct.v11i5.1142.

Di Martino, B., Li, K.-C., Yang, L. T., & Esposito, A. (2018). Trends and strategic researches in internet of everything. In B. Di Martino, K.-C. Li, L. T. Yang, & A. Esposito (Eds.), *Internet of everything: Algorithms, methodologies, technologies and perspectives* (pp. 1–12). Singapore: Springer Nature.

Dircke, P., & Molenaar, A. (2010). Smart climate change adaptation in Rotterdam, delta city of the future. *Water Practice Technology*, *5*(4). https://doi.org/10.2166/wpt.2010.083. wpt2010083.

Eriksson, M., Niitamo, A.-P., & Kulkki, S. (2005, January). *State-of-the-art in utilizing living labs approach to user-centric ICT innovation—A European approach (working paper)*. Sweden/ Helsinki, Finland: CDT at Lulea University of Technology, Nokia/Centre for Knowledge and Innovation Research at Helsinki School of Economics.

Garrigos-Simon, F. J., & Narangajavana, Y. (2015). From crowdsourcing to the use of masscapital. The common perspective of the success of Apple, Facebook, Google, Lego, TripAdvisor, and Zara. In F. J. Garrigos-Simon, I. Gil-Pehuan, & S. Estelles-Miguel (Eds.), *Advances in crowdsourcing* (pp. 1–13). Switzerland: Springer.

Gatautis, R., & Vitkauskaite, E. (2014). Crowdsourcing application in marketing activities. *Procedia-Social and Behavioral Sciences*, *110*, 1243–1250. https://doi.org/10.1016/j.sbspro.2013.12.971.

Giffinger, R., Fertner, G., Kramar, H., Kalasek, R., Pichler-Milanović, N., & Meijers, E. (2007). *Smart cities—Ranking of European medium-sized cities (final report)*. Vienna, Austria: Vienna University of Technology.

Glasmeier, A., & Christopherson, S. (2015). Thinking about smart cities. *Cambridge Journal of Regions, Economy and Society*, *8*(1), 3–12. https://doi.org/10.1093/cjres/rsu034.

Huang, H., Gartner, G., Krisp, J. M., Raubal, M., & Van de Weghe, N. (2018). Location based services: Ongoing evolution and research agenda. *Journal of Location Based Services*, *12*(2), 63–93. https://doi.org/10.1080/17489725.2018.1508763.

Huovila, A., Bosch, P., & Airaksinen, M. (2019). Comparative analysis of standardized indicators for smart sustainable cities: What indicators and standards to use and when? *Cities*, *89*, 141–153. https://doi.org/10.1016/j.cities.2019.01.029.

Hussain, A. M., & Hussain, M. M. (2015). CMOS-technology-enabled flexible and stretchable electronics for internet of everything applications. *Advanced Materials*, *28*(22), 4219–4249. https://doi.org/10.1002/adma.201504236.

Klopp, J. M., & Petretta, D. L. (2017). The urban sustainable development goal: Indicators, complexity and the politics of measuring cities. *Cities*, *63*, 92–97. https://doi.org/10.1016/j.cities.2016.12.019.

Koch, F., & Krellenberg, K. (2018). How to contextualize SDG 11? Looking at indicators for sustainable urban development in Germany. *International Journal of Geo-Information*, *7*(12), 464. https://doi.org/10.3390/ijgi7120464.

Letaifa, S. B. (2015). How to strategize SMART cities: Revealing the SMART model. *Journal of Business Research*, *68*(7), 1414–1419. https://doi.org/10.1016/j.jbusres.2015.01.024.

Li, S., Xu, L. D., & Zhao, S. (2015). The internet of things: A survey. *Information Systems Frontiers*, *17*, 243–259. https://doi.org/10.1007/s10796-014-9492-7.

Lombardi, P. (2011). New challenges in the evaluation of smart cities. *Network Industries Quarterly*, *13*(3), 8–10.

MacDonald, A., Clarke, A., Huang, L., Roseland, M., & May Seitanidi, M. (2018). Multi-stakeholder partnerships (SDG #17) as a means of achieving sustainable communities and cities (SDG #11). In W. Leal Filho (Ed.), *World sustainability series. Handbook of sustainability science and research* (pp. 139–209). Cham: Springer. https://doi.org/10.1007/978-3-319-63007-6_12.

Manville, C., Cochrane, G., Cave, J., Millard, J., Pederson, J. K., Thaarup, R. K., et al. (2014, January). *Mapping smart cities in the EU (study—Document requested by the European Parliament's committee on industry, research and energy)*. Brussels, Belgium: Directorate General for Internal Policies, Policy Department A: Economic and Scientific Policy.

Mitchell, S., Villa, N., Stewart-Weeks, M., & Lange, A. (2013). *The internet of everything for cities: Connecting people, process, data, and things to improve the 'livability' of cities and communities*. Cisco. Retrieved from: http://www.cisco.com/c/dam/en_us/solutions/industries/docs/gov/everything-for-cities.pdf.

Mora, H., Gilart-Iglesias, V., Pérez-del Hoyo, R., & Andújar-Montoya, M. D. (2017). A comprehensive system for monitoring urban accessibility in smart cities. *Sensors*, *17*(8), 1834. https://doi.org/10.3390/s17081834.

Mora, H., Gilart-Iglesias, V., Pérez-del Hoyo, R., Andújar-Montoya, M. D., & Compañ-Gabucio, H. J. (2016). Interactive cloud system for the analysis of accessibility in smart cities. *International Journal of Design & Nature and Ecodynamics*, *11*(3), 447–458. https://doi.org/10.2495/DNE-V11-N3-447-458.

Nilsson, M., Griggs, D., & Visbeck, M. (2016). Map the interactions between sustainable development goals. *Nature News*, *534*(7607), 320–322.

Papadopoulou, C.-A., & Giaoutzi, M. (2017). Crowdsourcing and living labs in support of smart cities' development. *International Journal of E-Planning Research*, *6*(2), 22–38. https://doi.org/10.4018/IJEPR.2017040102.

Papadopoulou, C.-A., & Hatzichristos, Th. (2020). Allocation of residential areas in smart insular communities: The case of Mykonos, Greece. *International Journal of E-Planning Research*, *9*(4), 40–60. https://doi.org/10.4018/IJEPR.2020100103.

Plageras, A. P., Psannis, K. E., Stergiou, C., Wang, H., & Gupta, B. B. (2018). Efficient IoT-based sensor BIG data collection—Processing and analysis in smart buildings. *Future Generation Computer Systems*, *82*, 349–357. https://doi.org/10.1016/j.future.2017.09.082.

Ramaprasad, A., Sánchez-Ortiz, A., & Syn, T. (2017). A unified definition of a smart city. In M. Janssen, K. Axelsson, O. Glassey, B. Klievink, R. Krimmer, I. Lindgren, et al. (Eds.), *Electronic government. EGOV 2017. Lecture notes in computer science, 10428* (pp. 13–24). Cham: Springer. https://doi.org/10.1007/978-3-319-64677-0_2.

Ray, P. P. (2018). A survey on internet of things architectures. *Journal of King Saud University—Computer and Information Sciences*, *30*(3), 291–319. https://doi.org/10.1016/j.jksuci.2016.10.003.

Riganti, P. (2017). Smart cities and heritage conservation: Developing a smartheritage agenda for sustainable inclusive communities. *ArchNet-IJAR*, *11*(3), 16–27. https://doi.org/10.26687/archnet-ijar.v11i3.1398.

Schaffers, H., Komninos, N., Pallot, M., Trousse, B., Nilsson, M., & Oliveira, A. (2011). Smart cities and the future internet: Towards cooperation frameworks for open innovation. In J. Domingue, et al. (Eds.), *The future internet, FIA 2011: Achievements and technological promises—Lecture notes in computer science* (pp. 431–446). Berlin, Heidelberg: Springer.

Shyam, G. K., Manvi, S. S., & Bharti, P. (2017). Smart waste management using internet-of-things (IoT). In *Proceedings of the 2nd international conference on computing and communications technologies (ICCCT), 23–24 February, Chennai, India*. https://doi.org/10.1109/ICCCT2.2017.7972276.

Thuan, N. H., Antunes, P., & Johnstone, D. (2016). Factors influencing the decision to crowdsource: A systematic literature review. *Information Systems Frontiers*, *18*(1), 47–68. https://doi.org/10.1007/s10796-015-9578-x.

United Nations. (2020). Retrieved from: https://www.un.org/sustainabledevelopment/sustainable-development-goals/.

Visvizi, A., & Lytras, M. (2018). Rescaling and refocusing smart cities research: From mega cities to smart villages. *Journal of Science and Technology Policy Management*, *9*(2), 134–145. https://doi.org/10.1108/JSTPM-02-2018-0020.

Visvizi, A., & Lytras, M. (2019). Smart cities research and debate: What is in there? In A. Visvizi, & M. Lytras (Eds.), *Smart cities: Issues and challenges—Mapping political, social and economic risks and threats* (pp. 1–14). Amsterdam, The Netherlands: Elsevier.

Voytenko, Y., McCormick, K., Evans, J., & Schliwa, G. (2016). Urban living labs for sustainability and low carbon cities in Europe: Towards a research agenda. *Journal of Cleaner Production, 123*, 45–54. https://doi.org/10.1016/j.jclepro.2015.08.053.

Weitz, N., Carlsen, H., Nilsson, M., & Skånberg, K. (2018). Towards systemic and contextual priority setting for implementing the 2030 agenda. *Sustainability Science, 13*, 531–548. https://doi.org/10.1007/s11625-017-0470-0.

Whitla, P. (2009). Crowdsourcing and its application in marketing activities. *Contemporary Management Research, 5*(1), 15–28. https://doi.org/10.7903/cmr.1145.

Yigitcanlar, T., Kamruzzaman, M., Foth, M., Sabatini-Marques, J., da Costa, E., & Ioppolo, G. (2019). Can cities become smart without being sustainable? A systematic review of the literature. *Sustainable Cities and Society, 45*, 348–365. https://doi.org/10.1016/j.scs.2018.11.033.

Part II

Why smart governance is a necessity for the attainment of the SDGs and how to make it truly smart? Involving and listening to the citizens

Chapter 5

Governance and SDGs in smart cities context

Shahira Assem Abdel-Razek
Delta University for Science and Technology, Department of Architectural Engineering, Dakahlia, Egypt

1. Governance in urban policy

Governance is defined as "the process of decision-making through which decisions are implemented (or not implemented)" (UN, 2020), while urban governance in one definition is "the process by which governments (local, regional, and national) and stakeholders collectively decide how to plan, finance, and manage urban areas" (Governance and Social Development Resource Centre, 2016).

The term "urban," currently, has evolved to encompass and bridge all structures related to the city; a complex meaning that binds together the cities of today and how they will evolve into those of tomorrow. The latest trends in the urban setting differ from one researcher to the other, some identify them as the bigger urban commemorates, while others define them as any city that is evolving. From digitalization (smart cities), green cities, blue corridors, innovation nodes, megacities, inclusive cities, self-producing cities, adaptive cities, mitigative cities, resilient cities, walkable cities, health-promoting cities (cities for well-being), net zero-waste cities, are all some of the current urban trends out there, and although there are clear overlaps between most of them and the probability that any one city may constitute several of these, the concept of starting any developmental projects with clear studies for (Lytras, Visvizi, Chopdar, Sarirete, & Alhalabi, 2021) cost-effectiveness and sustainability is now inevitable in order to be labeled sustainable.

Urban governance began as a transformation from governing to governance, focusing on a more holistic general approach to how to manage and govern an urban space through more than mere laws, rules and regulations, utilizing a participatory approach for the implementation of key parameters like transparency, subsidiarity, inclusivity, and responsivity (Bellamy, 2017; DiGaetano & Strom, 2003; Kearns & Paddison, 2000). Good urban governance does not focus only on research and analysis but also on transparency, accountability,

Smart Cities and the UN SDGs. https://doi.org/10.1016/B978-0-323-85151-0.00005-1

efficiency, public participation and mobility, equality, dispersion of authority, and social cohesion.

However, governance not only deals with the process and aftermath of a project's implementation but acts as a vital role in the initiation and decision-making phase also, from the start of the necessary studies that a project undergoes to be implemented and until the end-stage and turnover, governance plays a lead role in the process. This process, in essence, analyzes and influences the revenue behind economic growth for the underprivileged and determines how policies, institutional laws and regulations, and mechanisms may help evolve decisions and outcomes to help them. It is thus feasible to say that if this process may help achieve security and equality to those living in lower economies, it may certainly help and aid those at the multisectoral and project-based level to plan, analyze and improve outcomes of housing and other developmental projects.

This chapter will focus on governance in some of these cities, namely smart cities, walkable cities, cities for health, and well-being. The remainder of the chapter will discuss the importance of integrated research in urban governance, in an attempt, to formulate and determine a set of studies for urban policies to better aid urban development in reaching sustainability in new urban trend settings.

2. Methodology

The undertaken methodology is an analytical overview of governance policies and procedures present and their relationship with urban development projects and sustainability.

3. Urban planning research

In the field of urban development, governance is a concept embedded in the context of transparency. However, the information provided to the public regarding the analysis made for any development project and the benefit of the average citizen from this data is a somewhat gray area. In an ideal setting, the public should be made aware of the environmental, health, and financial risk of any project, as well as the possible benefits and revenue. This, in turn, will also incur accountability.

As an example, the construction industry in Egypt is an integral part of the economy and plays an important role in the GDP of the country: 7% in 2015 and 15% in 2017 (Selim, 1985; UKTI, 2020). The residential real estate sector, in Egypt today, is considered one of the most important investment sectors with a total number of vacant housing units estimated around 12.8 million units, 4.6 million of which are fully constructed and 4.3 million units still under construction. This leaves an approximate 2.8 million units closed, or due to be occupied (Central Agency for Public Mobilization and Statistics, 2016). With the boom in the construction sector over the past 3 years, more and more units are being put up, not to mention the new towns and cities being planned, this in itself is an integral reason of why developmental studies are needed before the construction of any new project if only to predict the cost-effectiveness of the funds allocated to be spent.

These past numbers, inevitably, lead to the speculation of effective planning regarding whether or not feasibility studies for these projects were undertaken and the strategies needed to rectify the situation. It is also important to highlight that a metering system does not exist for the price of land or for the average retail or lease price, an issue that leaves buyers and renters at the mercy of whoever is behind the project. It is regardless to say that to be able to unleash the full potential of any city, investing in its main components is a key player. These components include but are not limited to residential, commercial, and industrial projects, with direct availability to secondary services supported by effective markets, affordable accessibility, durable services, proper governing laws and regulations, appropriate funding and transparent and accountable city-level political systems. All of which are the key components of governance.

The planning of any project is subdivided into three phases; preparation, planning, and implementation, however, it is without doubt, that although evaluation helps shed light on the problems, hindrances, obstacles, and drawbacks of any project, the phase of evaluation is not always included in the overall planning, and that is where feasibility studies should be able to help. Feasibility studies are studies that are carried out at the early onset of large, complex, or developmental projects to assess its cost efficiency and point of revenue in a given time and place. This does not necessarily have to be only in cases where the project is of a skeptical nature or controversial. These studies should act as a tool to aid decision makers and stakeholders to reach a verdict that otherwise may not have been possible, thus establishing the viability of the project and determining program, design and building options and their impact on operational costs and allocated capital, while providing the needed data required for the rest of the projects' documents, as well as the plan and strategic brief (Shapers, 2020).

When dealing with large or complex projects, the studies carried out may differ from one type to the other. This is mainly due to the parameters and dimensions of each project and the expected outcome, however, it is safe to categorize these studies into four main types; environmental, demographic, technical, and financial studies (El Shimy, 2008).

At the onset of any project, the need for establishing it should be confirmed first, and several points need to be taken into consideration before the launch of any urban project, these include but are not limited to: a needs assessment for the local and surrounding community, time frame for allocated budget and expenditure to ensure efficiency and time till point of revenue is achieved, data analysis of existing surrounding infrastructure and services to diminish the possibility that the project may add an additional burden to the existing facilities, appropriate accessibility solutions for areas not connected to the grid to ensure continuity and connectivity with actual, present, appropriate and affordable transport, emissions strategy to mitigate and decrease possible emissions resulting from both construction and operating phases, study of the demographic status of the surrounding areas to ensure social acceptability to the context it is built in, possibility of new market and job opportunities for the local community to aid in their empowerment, the development of the surrounding areas to facilitate selling and sustainability of the project.

These points, among others, should be at the forefront for stakeholders, decision and policymakers, as well as, participants from the local community and should be addressed and answered. They should also be included in the project documents and formulated at the beginning of the planning phase of the project. These points do not deflect from the need to carry out the needed studies as mentioned before, however, the studies may differ from one type of urban setting to the other, with many of them being obligatory to all and some only to the specific type of urban city planned. As mentioned before, four distinct types of studies should be carried out: environmental, demographic, technical, and financial. The focus of each of these studies differs from one location to the next and from one urban setting to the other, however, the majority of urban projects need the following:

- Environmental studies
    - To focus on the impact of the immediate environment surrounding the project, assessment of factors, such as types and procurement places for building materials (proximity of procurement will affect the amount of fuel used and the amount of emissions exuded), life cycle assessments of building materials, fuel consumption and emissions, sanitation and water sources for the developing of the project, accessibility to site for working man power and labor (also included in fuel emissions), and types of energy used in the development process is needed.
- Demographic studies
    - To elaborate on the demographics of the project, the beneficiaries, (both in terms of man power and labor) surrounding equity in terms of who will buy into it, as well as past trends and forecasts of the rate of population increase and the average income of the public in that area.
- Technical studies
    - To include planning and construction permits and approvals, future expansion and development of the targeted area, assessment and on-site analysis of the project area, access to existing facilities and services (instead of building new ones), procurement options, and distances, specific program-related considerations.
- Financial studies
    - To reflect on proposed budgets and financial allocations (allocated funding in reference to project requirements), starting with the cost of site studies,

permits, to operational and maintenance costs. It also includes the cost of fuel for construction and transportation (for both materials and labor) and the cost of accommodating labor if the project is in a remote area. Financial studies should also be carried out to assess the rapidity of return on investment to assess whether the time is right for the project to be implemented or not. The above studies should also take into consideration inflation rates and financial risk management parameters. Financial studies should not only focus on the cost of construction but also on the idea of demolition, should the need arise for pressing country developments.

The previous detailed studies, as well as a full "Cradle-to-Grave cycle" should be addressed in all forms of studies; for example, the environmental impact of tearing down the project, as well as, the social burden and acceptance resulting from this decision.

4. Linking comprehensive sustainable research, SDGs, and urban development

With the adoption of the 17 post-2015 SDGs, a new challenge faced most researchers focusing on the parameters of their implementation, i.e., the concept of governance. The concepts of how to claim commitment to the SDGs through governance is best described, by many, as being merely theoretical (Boas, Biermann, & Kanie, 2016; Bowen et al., 2017; Kanie, Zondervan, & Stevens, 2014; Meuleman & Niestroy, 2015) and in dire need of a hands-on integrated and holistic strategy to intertwine the relations between the goals and their indicators to better achieve sustainability. Governance for the SDGs needs an appropriate and enabling environment to be able to function with decision and policymakers involved in the dynamics of planning and implementation of the strategies that cohere the different goals together. Governance is the fourth pillar of the sustainable development aside from the social, economic, and environmental aspects that was established in 2014 to ensure that the result of the intersection of the initial three pillars will be focused on and dealt with transparency, accountability, and subsidiarity, through active involvement of stakeholders, community participation, and other key players. It is thus safe to say, that without effective governance and appropriate advanced research, sustainability cannot be achieved. It is also futile to assume that governance alone will achieve the goals without taking into consideration all the comprehensive academic and theoretical approaches and the studies that go with it. In order to better focus on the development and practice of governance on the SDGs, it was mentioned earlier that this research would only focus on some of the evolving urban trends, including smart cities, and walkable healthy cities.

Ever since the smart cities movement started, it has focused on the emergence of the development of information and communication technologies. Since then, people have become increasingly dependent on these technologies, such as telecommunications, automation, surveillance, and security; in short, the Internet of Things (IoT). Smart cities utilize advanced solutions to optimize the quality of life through responsive governance (All, 2020). However, the amount of data imparted and the willingness of citizens to give this data away is somewhat a vague area, especially given the fact that consumers do not know the analyzer and beneficiaries of their data, as well as the extent of the usage of their data. The consumer market should have these analytics commonly known and this knowledge dispersed to ensure accountability and transparency to its customers, thus ensuring that governance is reached. Another point worth mentioning is the technical feasibility and dreams of smart cities today. It is no secret, that what was not possible in the near past is now an everyday occurrence, however, the cost of this "smart" know-how needs funding for it to become openly ready for public use. The idea behind conducting a set of comprehensive studies should be able to clear the path for the use of these technologies, meanwhile, a focus should be on users' interaction and participation within a set of smart city values and determinants. A void exists between the capabilities of smart cities and the research undertaken, mainly due to inappropriate infrastructure and both social and political hindrances, however, users themselves and the extent of their knowledge and use of smart services is what makes a city smart more than the actual presence of the technology (Visvizi & Lytras, 2018).

The focus on smart cities is one that is taken upon by crosscutting personnel and professionals in several fields, which together, adds up to the idea of integration. Urban and the built environment professionals have now become more accustomed to ideas like health and well-being, quality of life, entertainment, and leisure and recreation. Other sectors like education, employment, public participation, innovation, and entrepreneurship and philanthropism have also surfaced within the new vocabulary of urban settings and smart cities. The above denotes that consensual thought is imperative to achieve a holistic approach to smart cities with sustainability, governance, and resilience at its core.

4.1 Walkable cities-healthy cities

Reports as early as 1861 emphasize the importance of walking for the body, not only did it focus on its importance physically but also spiritually (Stitt, 2017). Walking helps in reducing risks of hypertension, metabolizes cholesterol levels, increases cardiovascular health, and aids and maintains weight loss. It also helps in alleviating the mood and reducing anxiety and stress levels (Carpenter, 2013; Morris & Hardman, 1997; Rippe, Ward, Porcari, & Freedson, 1988). However, ever since man has become accustomed to using a car, health has dwindled due to lack of exercise. The idea of reaching everywhere by transportation and being dependent on the transport systems is a luxury that has turned evil. Physical exercise is essential for the overall well-being of people. It reduces the incidence of heart disease and cardiac arrest, helps promoting mental health, aids in depression therapy, helps metabolism and aids in weight loss, and alleviates the mood. Walking as a form of exercise does all the above and contributes as a clean source of transportation from one place to the other. Needing a car to run simple errands, like going to the grocery store, or basic everyday life chores is an urban planning problem that needs solving. The link between walking and the importance of urban design is furthered with research on the effect of the built environment and the idea of being able to reach the specific destination by foot for both emotional and physical well-being (Corburn, 2004; Handy, Boarnet, Ewing, & Killingsworth, 2002; Jackson & Kochtitzky, 2010; Saelens & Handy, 2008).

The focus on urban health and healthy cities accentuates the use of the public health paradigm and methods to denote urban processes, systems, and solutions through studying the different demographic determinants available (Capolongo, Battistella, Buffoli, & Oppio, 2011). It also takes into consideration the systems and environment of the city, inevitably connecting these together to form a "health prism" model, which is sometimes referred to as "health lens" (Fehr & Capolongo, 2016). The key parameters of these models are promotion of health and prevention of disease and their interconnectivity with urban factors. The focus on urban health measures conforms to the WHO's "Health in all policies" concept, which discusses the connectivity of societal, environmental, planning and policy factors to bridge over the idea of conventional health paradigms, thus resulting in the contribution of professionals from all sectors of the community to aid in the decision-making process, an initial key factor in governance. The idea that urban health links healthcare professionals together with experts from several fields like city planners and designers, architects, landscape designers, environmental and construction engineers, fiscal agents, real estate and brokerage firms, and many others, is a true point to achieving the SDGs.

The health of citizens is of the top priorities for urban planners, developers, and designers, and across the past decade the idea of making cities "walkable" has been of concern (Speck, 2012, 2018). The benefits of a walkable city range from health and prosperity of living individuals to better economic attainment and better environmental control. The idea that transportation should not be made the core value around which to plan a city has become forefront in most planners' minds, that is, "Planning for humans instead of planning for vehicles." Walkability also affects overall citizenship through increased property value, better physical health, increased cognitive functions, lower crime rates, and higher civic engagement (Castro, 2020).

4.2 Smart cities vs walkable cities

As mentioned before, smart cities use sensors, applications, and IoT to gather data to improve the living experience and quality of life of its individuals, while aiding in their overall health and well-being. These parameters are not only linked to those of the smart city itself but also reflect on any other type of community therein. Using sensors in cameras, CCTVs, pedestrian lights, and even smart roads and sidewalks may help policy and officials making decision assess which areas need more pedestrian routes than others. Sensors and applications from specific shops and retail stores provide guidance as to the flow of clientele arriving and the need for improvement. These metrics along with walking applications help gather data as to which areas need more attention than others and if the planning and design of the city need development. These applications should be made aware to the citizens to help them make an informed decision on whether or not they agree to send feedback and metrics to these sensors, or even, have their steps counted and analyzed (governance). The main hindrance here is that most of this data may fall under national security and may be at risk of being hacked. These metrics should also be linked with the health records of individuals for those at risk of hypertension or strokes to enable faster reach should a health crisis arise, and a physician is needed.

To further demonstrate the link between smart cities and walkable cities, it can be argued that global urban health can be improved through building smart cities, this can be achieved through smarter transportation, smarter infrastructure, services and utilities, better-equipped lighting, heating and ventilation networks that are embedded with optimum technology (Visvizi, Mazzucelli, & Lytras, 2017).

5. Conclusion

In conclusion, urban governance is a set of indicators, processes, and evaluators governed together to achieve urban sustainability. These indicators are derived from the different players that are the main constituents of the urban scape industry but are not limited in importance to urban projects. Resilience, sustainability, longevity, practicability, cost-effectiveness, and efficiency, are some of these players and can only be achieved with a set of comprehensive research and appropriate indicators and studies. Without which many developmental projects may well turn out to be futile. The key concept of a well-balanced holistic project is sustainability and the steps that pave the way to achieve it. The sustainable development goals are wholesome if applied and taken well into consideration at the planning level. However, monitoring and evaluation are of the same importance, if not more, to ensure that the project was well planned and carried out in an appropriate manner. Having this extensive research broadcasted to the public as well as opening channels of participation for them will lead to better benefits and much-needed governance. Smart cities, walkable cities, and many more new urban settings are optimum opportunities to strategize, plan, implement, and evaluate urban governance. These cannot only be termed feasibility studies as they encroach much more than what traditional and conventional feasibility studies today encompass, but they may build up on the idea of the research that the main aspects involved are environmental, demographic, technical, and financial, thus ensuring the transparency of the urban project and eventually urban governance.

References

All, I. F. (2020, June 18). www.iotforall.com/what-makes-smart-city-2019. Retrieved from: www.iotforall.com.

Bellamy, R. (2017). From government to governance. In *From government to governance* (1st ed.). Routledge. https://doi.org/10.4324/9781315254920.

Boas, I., Biermann, F., & Kanie, N. (2016). Cross-sectoral strategies in global sustainability governance: Towards a nexus approach. *International Environmental Agreements*, *16*(3), 449–464. https://doi.org/10.1007/s10784-016-9321-1.

Bowen, K. J., Cradock-Henry, N. A., Koch, F., Patterson, J., Häyhä, T., Vogt, J., et al. (2017). Implementing the "sustainable development goals": Towards addressing three key governance challenges—Collective action, trade-offs, and accountability. *Current Opinion in Environmental Sustainability*, *26–27*(February), 90–96. https://doi.org/10.1016/j.cosust.2017.05.002.

Central Agency for Public Mobilization and Statistics (CAPMAS). (2016). *Census official data*. Cairo: Governmental Body.

Capolongo, S., Battistella, A., Buffoli, M., & Oppio, A. (2011). Healthy design for sustainable communities. *Annali di Igiene: Medicina Preventiva e di Comunita*, *23*(1), 43–53.

Carpenter, M. (2013). From "healthful exercise" to "nature on prescription": The politics of urban green spaces and walking for health. *Landscape and Urban Planning*, *118*, 120–127. https://doi.org/10.1016/j.landurbplan.2013.02.009.

Castro, D. (2020, June 22). *Govtech*. Retrieved from: https://www.govtech.com/analytics.

Corburn, J. (2004). Reconnecting urban planning and public health. *American Journal of Public Health*, *94*(4), 541.

DiGaetano, A., & Strom, E. (2003). Comparative urban governance: An integrated approach. *Urban Affairs Review*, *38*(3), 356–395. https://doi.org/10.1177/1078087402238806.

El Shimy, H. (2008). Integral feasibility studies ae the tools of ruling continual development in the new urban societies in Egypt. In *Ain Shams international conference for architecture & urban planning* (pp. 1–16). Cairo: Ain Sham University.

Fehr, R., & Capolongo, S. (2016). Healing environment and urban health. *Epidemiologia e Prevenzione*, *40*(3–4), 151–152. https://doi.org/10.19191/EP16.3-4.P151.080.

Governance and Social Development Resource Centre (GSDRC). (2016). *Urban governance*. Birmingham: DFID Crown.

Handy, S. L., Boarnet, M. G., Ewing, R., & Killingsworth, R. E. (2002). How the built environment affects physical activity: Views from urban planning. *American Journal of Preventive Medicine*, *23*(2 Suppl), 64–73. https://doi.org/10.1016/s0749-3797(02)00475-0.

Jackson, R. J., & Kochtitzky, C. (2010). *Creating a healthy environment: The impact of the built environment on public health*. Director, October (p. 20).

Kanie, N., Zondervan, R., & Stevens, C. (2014). Ideas on governance "of" and "for" sustainable development goals. In *UNU-IAS/POST2015 conference report*.

Kearns, A., & Paddison, R. (2000). New challenges for urban governance. *Urban Studies*, *37*(6), 845–850.

Lytras, M. D., Visvizi, A., Chopdar, P. K., Sarirete, A., & Alhalabi, W. (2021). Information Management in Smart Cities: Turning end users' views into multi-item scale development, validation, and policy-making recommendations. *International Journal of Information Management*, *56*, 102146. https://doi.org/10.1016/j.ijinfomgt.2020.102146.

Meuleman, L., & Niestroy, I. (2015). Common but differentiated governance: A metagovernance approach to make the SDGs work. *Sustainability*, *7*(9), 12295–12321. https://doi.org/10.3390/su70912295.

Morris, J. N., & Hardman, A. E. (1997). Walking to health. *Sports Medicine*, *23*(5), 306–332. https://doi.org/10.2165/00007256-199723050-00004.

Rippe, J. M., Ward, A., Porcari, J. P., & Freedson, P. S. (1988). Walking for health and fitness. *JAMA*, *259*(18), 2720–2724. https://doi.org/10.1001/jama.1988.03720180046031.

Saelens, B. E., & Handy, S. L. (2008). Built environment correlates of walking: A review. *Medicine and Science in Sports and Exercise*, *40*(7 Suppl), 550–566. https://doi.org/10.1249/MSS.0b013e31817c67a4.

Selim, T. M. (1985). Development of the construction industry in Egypt. *International Journal of Project Management*, *3*(1), 27–34.

Space Shapers. (2020, June 14). Retrieved from: https://www.spaceshapersarchitects.com/feasibility-report-building-construction/.

Speck, J. (2012). *Walkable City: How downtown can save America one step at a time*. New York: North Point Press.

Speck, J. (2018). *Walkable City rules: 101 steps to making better places*. Washington, DC: Island Press.

Stitt, J. (2017, October 11). *A short history of walking*. Garrison Institute. Retrieved from: https://www.garrisoninstitute.org/short-history-of-walking.

UKTI. (2020, June 18). *Open to export*. Retrieved from: https://opentoexport.com.

UN. (2020, June 12). *What is good governance*. Retrieved from: United Nations Economic and Social Comission for Asia and the Pacific (ESCAP). https://www.unescap.org/sites/default/files/good-governance.pdf.

Visvizi, A., & Lytras, M. D. (2018). Rescaling and refocusing smart cities research: From mega cities to smart villages. *Journal of Science and Technology Policy Management*, *9*(2), 134–145. https://doi.org/10.1108/JSTPM-02-2018-0020.

Visvizi, A., Mazzucelli, C., & Lytras, M. (2017). Irregular migratory flows: Towards an ICTs' enabled integrated framework for resilient urban systems. *Journal of Science and Technology Policy Management*, *8*(2), 227–242. https://doi.org/10.1108/JSTPM-05-2017-0020.

Chapter 6

Reinterpreting governance in smart cities: An ecosystem-based view

Francesco Polese[a], Orlando Troisi[a], Mara Grimaldi[a], and Francesca Loia[b]
[a]*Department of Management & Innovation Systems, University of Salerno, Fisciano, Italy,*
[b]*Department of Economics, Management and Institutions University of Naples 'Federico II', Naples, Italy*

1. Introduction

The role of new technologies as key levers for the development of urban areas and the improvement of cities management leads extant research to conceptualize smart cities. The concept refers to the instrumented, interconnected, and intelligent (Harrison et al., 2010) urban areas that can pursue shared growth through an integrated set of technologies that shapes the interactions among actors (Nam & Pardo, 2011).

The need to address social challenges in smart cities is widely acknowledged in the literature (Höjer & Wangel, 2015; Visvizi & Lytras, 2018). Due to the different kinds of stakeholder groups with different needs which lie in urban areas, smart cities and smart regions should focus on the development of new sustainable and socially aware polices based on ICTs (*information & communication technologies*)-enabled solutions. Therefore smart technologies can foster the emergence of sustainability and innovation, which are two key levers to tackle social challenge for the improvement of cities' well-being (Visvizi, Lytras, Damiani, & Mathkour, 2018).

Sustainable cities and communities are relevant issues in the United Nations "Sustainable Development Goals" (SDGs), the key strategic principles included in the 2030 Agenda for Sustainable Development, which considers urban well-being as one of the 17 main social challenges established in the UN program. In detail, the 11th goal of SDGs identifies the necessity for the enhancement of urban quality of life (from sustainable transport systems, urbanization, waste management, etc.) and for the creation of sustainable and resilient cities (Goal 11).

Smart Cities and the UN SDGs. https://doi.org/10.1016/B978-0-323-85151-0.00006-3

Despite the recognition of the revolutionary role of technology in the growth of cities, the use of smart tools does not automatically allow the attainment of cities' well-being and innovation, since human and relational capabilities are needed. Smart technology requires necessarily entrepreneurial, managerial, and governance skills to foster the emergence of sustainable growth. Previous studies on smart cities do not introduce a unifying framework that explores different variables and strategic drivers (Appio, Lima, & Paroutis, 2019) for city's development (Lim, Kim, & Maglio, 2018). Moreover, the most recent advancements in the literature highlight the need to explore how entrepreneurship and governance can help to manage the criticalities in the administration of smart cities (Kitchin, 2015) and to improve policy-making for the attainment of sustainability (Visvizi et al., 2018).

For these reasons, to bridge the gap in the identification of the main drivers for a wise use of technology and of the key management and governance steps to pursue sustainability, *service ecosystem view* (Chandler & Vargo, 2011) is selected to reread smart cities as complex systems in which multilevel governance is required (Spohrer, Piciocchi, & Bassano, 2012; Vargo & Lusch, 2011).

The different levels of analysis formalized in ecosystems view allow to investigate how sustainability can be generated across the social (macro), relational (meso), and technological and human (micro) contexts in smart cities.

Therefore the study rereads the Italian municipality of Trento as a smart city according to an ecosystem-based view that explores the different actors, technology, resources, and cocreation practices (key ecosystem dimensions) in the city. Moreover, the strategies that governance should apply to harmonize multistakeholders' objectives and sustainable innovation are investigated. Empirical research seeks to address the following research questions:

RQ1: *Which are the main ecosystem's elements (actors, resources, technologies, institutions, and value propositions) in smart cities?*
RQ2: *How can governance harmonize the combination of these elements to pursue sustainable innovation?*

After the identification of the main ecosystem's elements involved in TrentoSmartCity, the ways in which multilevel governance can manage strategically these elements to attain social growth and different kinds of innovation at technological, cultural, and social levels are investigated.

The findings allow the proposition of a framework in which useful steps are revealed to help governance to understand how smart cities' decision-making and policy-making processes can be managed across multileveled contexts fostering social and economic development over time.

The chapter is structured as follows: in the first section, a literature review on the definition and key areas of smart cities is performed in order to emphasize the need to adopt an integrated view based on multilevel governance and service ecosystems. Then the findings from the case study on Trento Municipality are discussed to lay the foundations for the development of a theoretical framework that identifies the main drivers for sustainable innovation in smart cities. Finally, the conclusion, implications, and limitations of the study are elaborated.

2. Theoretical background

2.1 Smart cities

Smart cities are embedded systems based on sensors and interactive media that support knowledge diffusion and interaction. The success of smart cities is based on the integration of the three dimensions of urban space, technological, human, and institutional (Nam & Pardo, 2011), also defined as physical, institutional, and digital dimensions (Komninos, 2013; Visvizi & Lytras, 2018).

It is acknowledged that ICT can improve sustainability and quality of life by enhancing social capital, citizens' knowledge and fostering innovation (Stratigea, 2012). However, technology is conceived as one of the dimensions of a complex process in which human skills and intervention and the compliance with rules and social norms are essential for a proper application of technological tools.

In line with the need to adopt a systems view, the analysis of smart cities should not imply (only) the exploration of what a smart city is, but "a city's effort to make itself smart" (Nam & Pardo, 2011, p. 186). According to the literature (Caragliu, Del Bo, & Nijkamp, 2011; Nam & Pardo, 2011), the main drivers for smartness are: (1) technology; (2) human component; (3) institutional dimensions. The three factors are strictly interconnected: then a smart city invests in human, social capital and IT infrastructures to attain sustainable growth and improve the quality of life of actors/citizens through participatory governance (Polese et al., 2018).

Therefore the availability and quality of ICT infrastructure is a necessary but not sufficient condition for the attainment of smartness and development. As Giffinger, Fertner, Kramar, Meijers, and Pichler Milanovic (2007) emphasize, smart people who encourage mutual learning, social and ethnic plurality, creativity, open-mindedness, and participation in public life are an essential lever for the expansion of smart cities. According to Malek (2009), humanity—which represents cognitive, creative, and human abilities—makes cities smarter.

To examine the "human side" of smartness, the most recent contributions in the literature focus on the definition of the key role of governance in determining economic growth, innovation, and social inclusion (Visvizi et al., 2018). There is the need to detect new sustainable and socially aware polices for smart cities and to adopt a systems view that investigates the role of the social, cultural, and relational context in strategies and policies (Visvizi & Lytras, 2019). Thanks to the engagement of a series of actors (citizens, governmental bodies, public administrations, profit and nonprofit), innovation solutions can be codeveloped through a collaborative approach to meet relevant societal challenges (Lytras, Raghavan, & Damiani, 2017).

Based on the challenging complexity of markets and on the evolving requests of stakeholders, the need to detect the main drivers to foster the attainment of sustainable development is introduced (Visvizi et al., 2018). Thus, the social and environmental implications of the use of smart technologies in smart urban areas should be explored to identify the most proper strategies for governance and policy-making.

Despite the increasing interest in the analysis of smart cities according to a broadened view (Lytras & Visvizi, 2018), there is still the need to overcome the technology-oriented focus to explore the "social" side of innovation and the main enabling factors of sustainable outcomes enabled by smart people (Kourtit, Nijkamp, & Arribas-Bel, 2012).

Starting from the work of Paroutis, Bennett, and Heracleous (2014), an all-encompassing perspective on smart cities can be conceptualized thanks to the definition of three levels of analysis (depicted in Fig. 1): (1) *technology view* (ICT organizations that provide technological solutions); (2) *strategic view*; (3) *systems view* (the entire city and all the systems that implement the smart solutions). Firstly, the examination of city technology is needed to identify the actors who create and spread these technologies. Secondly, a strategic vision of city technology is needed to detect the processes and practices that organizations should implement to employ technology successfully. Lastly, the way in which the use of technology can foster the well-being of the whole smart city should be analyzed.

For this reason, in this study smart cities are reframed as smart places in which the technological dimension should be integrated necessarily with the institutional sphere of governance and the human abilities of management and policy-makers to harmonize the different stakeholders' finalities and align them with the overall cities' goals.

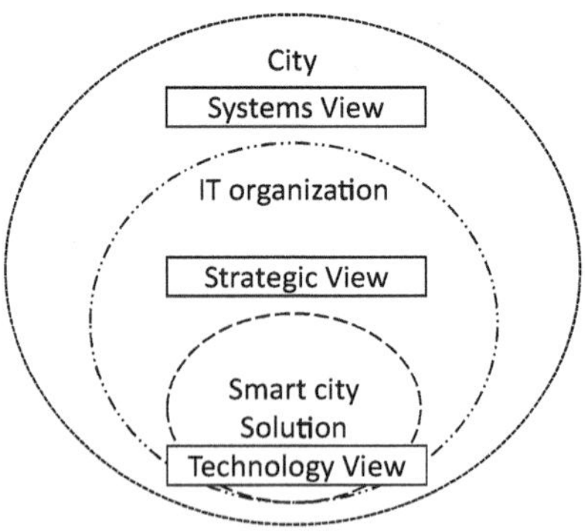

FIG. 1 Smart cities: three perspectives. *(Modified from Paroutis, S., Bennett, M., & Heracleous, L. (2014). A strategic view on smart city technology: The case of IBM smarter cities during a recession. Technological Forecasting and Social Change, 89, 270.)*

2.2 Governance in smart cities

A favorable institutional environment based on government support and on the successful design of governance policies is fundamental for the implementation of smart city initiatives.

In 2010 IBM conceptualized smart government as a key component of smart cities. Smarter government (Mellouli, Luna-Reyes, & Zhang, 2014) implies collaboration between various departments and with communities to become more effective and to provide citizens with access to information about decisions that affect their lives. The so-called "e-governance" is one of the most relevant enablers of citizens' participation in smart cities and of the enhancement of transparency in the decision and implementation process. A successful smart city can consist of top-down or bottom-up government approaches, but the active involvement of the different stakeholders is essential (Palmisano, 2008).

Thus innovation can be codeveloped thanks to citizens' empowerment and thanks to transparent decision-making processes that can support management's skills to turn information into knowledge. Paroutis et al. (2014) identify a strategic view on innovation that is redefined as a "users-driven" process generated bottom-up by actors with adequate digital knowledge promoted through social inclusion thanks to the establishment of open innovation business models (Marsal-Llacuna, Colomer-Llinàs, & Meléndez-Frigola, 2015). Hence, the smart use of technological tools depends on human's ability to exchange information through these tools and on smart governance (Pereira, Cunha, Lampoltshammer, Parycek, & Testa, 2017; Vanolo, 2016), which maximizes innovative and creative potential (Kitchin, 2015).

Smart governance (Lombardi, Giordano, Farouh, & Yousef, 2012) is intended as the overarching entity that can harmonize the different relationships among actors, technologies, and information flows in the cities to pursue innovation in the long run. The governance-oriented approach (Albino, Berardi, & Dangelico, 2015) stresses the need to adopt long-term strategies to manage and optimize the use of technology (Kitchin, 2015). Thus it is only through strategic governmental efforts to speed up bureaucratic processes that smart cities can create new solutions to improve sustainability and manage complexity (Stratigea, 2012; Toppeta, 2010). For this reason, due to the multidimensional nature of stakeholders' interests, needs, and relationships, multilevel governance seems to be the most adequate solution for smart cities (Gil-Garcia, Zhang, & Puron-Cid, 2016). Multilevel governance stems from the studies on networked governance (Kooiman, Bavinck, Chuenpagdee, Mahon, & Pullin, 2008; Stoker, 2006) and on systems view (Barile, Saviano, Polese, & Di Nauta, 2013; Piciocchi, Spohrer, Bassano, & Giuiusa, 2012) and conceptualizes the need to adopt diffused decision-making grounded on collaborative and participatory assets of power for the exploitation of innovation opportunities.

The implementation of multidimensional governance models requires the adoption of a systems and metalevel of analysis that explores the way in which the elaboration of different strategies for different actors, resources, and users' engagement practices can be harmonized. Hence, in the following paragraph, the service ecosystems view will be reframed as a suitable perspective for the systems analysis of smart cities.

2.3 Ecosystems view

The service ecosystems view (Akaka, Vargo, & Lusch, 2013; Chandler & Vargo, 2011) defines contemporary organizations as service systems embedded in multilevel contexts of exchange and resource integration (Storbacka, Brodie, Böhmann, Maglio, & Nenonen, 2016). The theory can be applied to reframe smart cities thanks to the adoption of a multilevel perspective that envelops actors' cocreation along the micro, meso, and macro contexts of analysis, which can generate social, political-institutional paths of shared growth. Then the ecosystems view allows a multidimensional analysis of the community's actors, purposes, cocreation mechanisms and of the potential different strategies of resource integration promoting the overall system's/city's growth.

Service ecosystems are "relatively self-contained, self-adjusting system[s] of resource-integrating actors connected by shared institutional logics and mutual value creation through service exchange" (Vargo & Lusch, 2011, p. 176). According to the most relevant contributions introduced in Service-dominant logic (Akaka et al., 2013; Barile, Ciasullo, Troisi, & Sarno, 2017; Storbacka et al., 2016; Vargo & Lusch, 2016), the key elements of ecosystems are: (1) *actors*: the generic "users" reframed as value cocreators; (2) *institutions*: coordination mechanisms (norms, social rules, practices) to attain shared purposes; (3) *value propositions*: the proposal and the sharing of common values; (4) *resource integration*: value and knowledge combination between and among actors; (5) *technology*: set of enabling tools that improves interactions and engagement.

The ecosystem's elements are embedded across multiple contexts: (1) micro (interactions); (2) meso (social networks of relationships); (3) macro (broader institutional context). Therefore the ecosystems view helps in identifying the actors, the technology, and the resources exchanged in smart cities and allows to shed light on the main features of a successfully diffused and multilevel governance for the attainment of growth and sustainable innovation.

To reread smart cities according to the ecosystems view, in Fig. 2 the main levels of analysis (technology, strategic, and systems view) identified by Paroutis et al. (2014) can be reread through the three contexts of the ecosystems view. Then cities can be reframed as ecosystems in which technology-mediated interactions can foster resource integration and value cocreation processes at the micro (design of a technological architecture), meso (strategic management of relationships), and macro levels (overall system's well-being). This multidimensional process needs to be harmonized by governance at the different ecosystem levels.

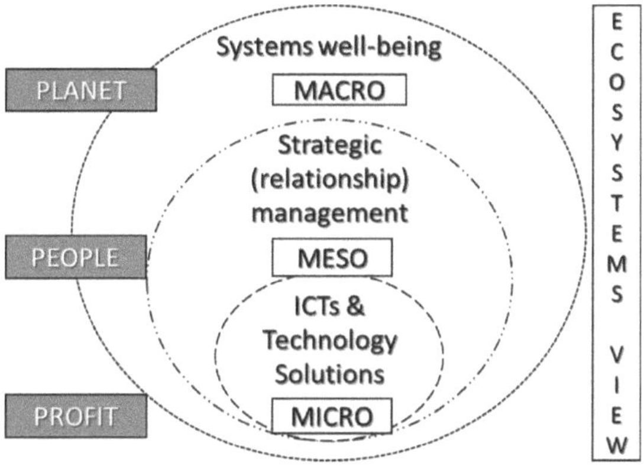

FIG. 2 Redefining the dimensions of sustainability in smart cities through the ecosystems view. *(Data from Author's elaboration.)*

Moreover, each ecosystem's context can be associated with three dimensions of the Triple Bottom Line (TBL) introduced by Elkington (1998)—profit, people, and planet—which can be associated, in turn, with the three dimensions of sustainability in smart cities (Bătăgan, 2011), which are economic (micro), social (meso), and political dimensions (macro). Different strategic drivers for sustainable development and different sustainable outcomes can be generated in the three levels, related to product, technological, and process improvement at the micro-level, to social and relational enrichment at the meso-level, to institutional and environmental growth at the macro-level.

3. Methodology

Case study methodology is employed to explore the Italian smart city of Trento (Northern Italy) as an ecosystem in which the main actors, technologies, resources, institutions, and the related value cocreation practices and sustainable innovation outcomes (at the economic, social, political/environmental level) can be explored along the three contexts of analysis defined in the theoretical background (micro, meso, macro).

Over the last 5 years, the municipality of Trento launched a set of technology-based projects and smart city initiatives, which enable the creation of an interconnected technological infrastructure involving citizens, government bodies, and private and public organizations.

Case study methodology (Yin, 1994) allows an in-depth investigation of the dynamics underlying a complex context according to a particularistic standpoint (Eisenhardt, 1989; Tellis, 1997). For this reason, it seems to be suitable for a careful investigation of multilevel units of analysis such as smart cities. To address the research questions, a qualitative content analysis based on secondary data is performed. In the exploratory stage of empirical research, a qualitative approach contributes to a richer analysis of phenomena referred to as multidimensional constructs that are not measured univocally and defined semantically in extant research. The technique employed, content analysis, allows to extract from texts (the unit of analysis) fewer content categories and permits to detect the focal points (Krippendorff, 2004) through the adoption of semantic criteria established by the researcher.

The main sources employed to collect data are: TrentoSmartCity official website; the official website of the municipality; the specific websites of the single smart projects; the official documents available in the different websites such as "Strategic plan," "Smart vademecum," "Guidelines paper"; Trento Municipality and Trento SmartCity social network pages.

The content analysis is based on questions administered ideally to the texts, obtained from a re-elaboration of the ecosystem's variables identified and defined in paragraph 2.3, and classified in Table 1. For each ecosystem's element (macro-dimensions), a specific subdimension is obtained from the preliminary application of the ecosystems view to the analysis of the city. Then a set of keywords is elaborated to detect more detailed features of the ecosystem that surrounds TrentoSmartCity.

Through a process of substruction (Dulock & Holzemer, 1991), the set of keywords, employed as guidelines to guide the first reading of the texts, are re-elaborated continuously to adapt to the results emerging while performing the process. Findings are obtained from a semantic interpretation of the texts from which the macro-categories identified by researchers are reconnected to the main variables of the research and classified to pinpoint the ecosystem's elements in the micro, meso, and macro-contexts.

TABLE 1 **The macro-dimensions, subdimensions, and keywords for content analysis.**

Macro-dimensions of analysis	Subdimensions	Keywords
Value propositions	Key values established, diffused, shared and internalized by management, governance, and actors	Ethics Environmental attitude Community culture Shared goals Mind-set Codesign Common values
Actors	Stakeholders engaged in the design, development and realization of the smart city projects	Citizens Institutions Organizations Nonprofit associations Engagement Joint policy-making Codevelopment

TABLE 1 The macro-dimensions, subdimensions, and keywords for content analysis—cont'd

Macro-dimensions of analysis	Subdimensions	Keywords
Technology	ICT, IT tools, analytics, machine learning and all the softwares and applications that support the smart city projects	Information and communication technologies Analytics Platforms E-government platforms Social networks Mobile applications Cloud computing Machine learning Artificial intelligence
Resources	Economic, material, informational, relational and social capital exchanged between and among actors in the city	Economic capital Material equipment Territorial assets Heritage Mind-set, attitude Knowledge Competencies Management abilities Relationship management
Institutions	Legal, political, economic, cultural and social norms-prerequisites for successful implementation of projects, value cocreation and sustainable innovation outcomes (new social/environmental standards, practices, and rules)	Legal standards Economic standards Cultural habits Social rules Environmental standards Social practices

Data from Author's elaboration.

4. Results

4.1 RQ1: Reinterpretation of TrentoSmartCity as an ecosystem

4.1.1 Value propositions

Value proposition at the micro-level is related to the purpose of spreading a cohesive culture throughout the city of Trento which aims at orienting the mind-set of citizens toward social change and at diffusing the ethical values related to concept of responsiveness toward sustainability. As reported in the mission on the official website of TrentoSmartCity, the smart projects implemented aim strategically at reducing the gap between political and economic interests and community well-being. Moreover, the goal of TrentoSmartCity is to challenge the barriers to innovation by realizing a "smart" use of technology and promoting digital knowledge among all the segments of the population, from young people to the elderly.

At the meso-level, the active sharing of the smart values established at the micro-level is carried out to create a cohesive culture that permeates the community. The goals are to enhance: (1) users' acceptance of technology; (2) users' attitude toward technology and willingness to participate in decision-making; (3) citizens' digital knowledge to make technologies become a "natural" component of their lives.

At the macro-level, the values internalized at the meso-level are actualized through the inclusion of citizens in decision-making, which can be translated into their involvement in the development of policy-making at a broader contextual level to solve issues related to mobility, energy, waste reduction, and general well-being. The codevelopment of policies and actions—derived from the decisions cocreated at the meso-level—is realized through projects that aim at pursuing digital citizenship enhancing users' commitment (e.g., "Welive," "Stardust," and "Digital Citizen").

4.1.2 Actors

The implementation of a general smart ecosystem that can empower the enactment of a set of smart technologies and initiatives surrounding TrentoSmartCity is supported by the following set of stakeholders actively engaged in the strategic codesign of a smart architecture: (1) businesses; (2) government bodies and institutions; (3) University and Research Institutes; (4) nonprofit associations and consortia. Trento municipality and the entire region own a network of businesses based on an innovation-oriented mind-set, proactiveness, and high flexibility in the labor system. Government bodies and institutions support actively digital innovation by providing funds and a local plan to promote smartness in the region. Moreover, the education system (Universities and Research centers) seems to be a key player in the development of a "smart place" with the introduction of degree and PhD courses based on the topic of smart technologies and sustainability.

The meso-context refers to the "step" in which actors are involved in service development and decision-making actively proposing innovative ideas and solutions for service inefficiencies. The stakeholders engaged in decision-making are: (1) citizens-students; (2) schools; (3) companies; (4) public administration; (5) research centers. Cocreated partnerships with schools and Universities can contribute to spreading knowledge on the use of smart technologies and to orient future generations toward the adoption of a sustainable mentality. Companies and public administration are key actors in codesign activities, playing a leading role in the transmission of knowledge and in the search for continuous improvement. Lastly, research centers provide data analysis and knowledge management skills to explore data collected from open platforms and to teach private and public employees to interpret research results.

The transition from meso-level to macro-level allows the "transformation" of users/decision-makers into real policy-makers. TrentoSmartCity organizes a set of projects that introduces an open government model to address environmental challenges and that advances a series of tools supporting citizens in their dialog with the municipality to redesign digitally both internal processes and government practices (e.g., "WeGovernment," "LosDama"). Citizens are also involved actively in the realization of sustainable actions (e.g., "Stardust").

4.1.3 Technology

TrentoSmartCity employs a harmonized set of technologies to spread and share smart culture with all the stakeholders involved, to improve existing services, and to provide new services for the (co-) creation of added value. "Decenter" is the most relevant project organized at the micro-level aimed at creating a general architecture that can foster the delivery of services and at promoting their efficiency and effectiveness. This project introduces a general computing platform based on a blockchain system to reach secure processing, automated operation, and timely delivery of services. Thanks to this, the provision of smart services for mobility, environment, and e-government is enabled.

At the meso-level, the main ICTs and smart tools adopted to implement the smart projects are machine learning and artificial intelligence instruments, which allow data collection in real time, enhance service recovery, and encourage users to propose new solutions. An integrated set of smart tools, such as networking technologies (distributed sensor systems, wireless, mobile platforms), gamification-based technologies, and eService Monitoring component, is employed to analyze user's data. Moreover, social networks are used to share news about the events and to extract relevant data through API, web scraping, content management systems, and clustering. Moreover, some mobile applications have been implemented, such as "APP 100% Riciclo" for waste disposal, "App Shelly" and "AppMyCicero" for mobility real-time information, and "TrentOrienta," codeveloped with students and based on general communications on the events in the city.

The technologies employed at the macro-level are: (1) open platforms (e.g., "Stardust") that permit citizens to actively contribute to the sustainable evolution of the city; (2) online informational desks to make citizens active participants in policy-making and to collect their feedback for the ongoing improvement of service (e.g., "Digital Citizen"). Users can propose ideas for new smart sustainable services (electric vehicles and mobile charging system, smart points, public lighting system). Moreover, the installation of a sensor network allows the collection of data on the environment, energy, mobility, safety, and waste disposal. The information extracted are shared in real time with the city through a digital dashboard.

4.1.4 Resources

At the micro-level, as described in the report of IEEE and in the ranking "ICityRate 2013," TrentoSmartCity owns a series of integrated resources that sustain the realization of a cohesive smart community. These are based on territorial assets that perform a sustainable use of resources, promote a high quality of life, and are based on a good infrastructure network to enable the use of technologies. The culture of smart city as a place is entrenched with a smart mind-set that all-encompasses the social and economic sphere. Local institutions share a philosophy based on justice, legality, and transparency. Profit and nonprofit organizations promote actively the enhancement of the labor system through the creation of new job positions and by exercising an innovation-oriented entrepreneurship. Lastly, local educational institutions seem to be devoted to the spreading of digital education culture throughout the community, from pupils to students to citizens.

The main resources provided at the meso-level are related mainly to technological knowledge and competencies (internal to private businesses or acquired from new data analysts or big data managers). Thus these skills are shared with citizens, students, public employees through educational activities. In addition, relational skills are exchanged in the community to enable more transparent communications and share experiences through smart technologies.

The key resources exchanged and cocreated at the macro-level are institutional skills and knowledge shaping a smart sustainable culture. Institutions and governmental bodies share knowledge and competencies about the use and the delivery of public services. The values of sustainability are internalized by citizens who provide immaterial skills deriving from a sustainable-oriented and innovation-oriented culture. Moreover, social skills are activated in the community network to enhance shared responsiveness toward sustainability paving the way for their commitment.

4.1.5 Institutions

The creation of a network of partnerships allows the diffusion of new legal and economic institutions. The emergence of this ecosystem allows to introducing new kinds of entrepreneurial and business models (with the introduction of new job positions such as big data managers) and new legal standards for digital rights, intellectual property, and delivery of services.

At the meso-level, institutions are related to the educational and social dimensions. In the first case, new participatory forms of culture and knowledge exchange and new ways of learning and teaching are created thanks to the organization of Living and Innovation Labs. In the second case, new forms of sociality introduced are: (1) ways of communicating through ICTs (more transparently); (2) forms of participatory (e-) decision-making; (3) rituals, such as the annual organization of "TrentoSmartCity week," that pave the way for new habits for the community.

Lastly, new institutions at the macro-level are formal and informal rules deriving from the introduction of: (1) forms of participatory government and open innovation business model for e-democracy; (2) ways of experiencing public services and relationship modalities, for example, citizens-government; (3) ways of inhabiting communities and of the living environment (environmental standards and new rules on mobility and environment, such as licensing for electric cars, car sharing).

4.2 RQ2: TrentoSmartCity governance

At a micro-level, TrentoSmartCity governance enables the creation of a set of values based on smart culture through the diffusion of an entrepreneurial attitude oriented toward the exploitation of innovation and sustainable development.

The strategic aim of TrentoSmartCity is related to the diffusion of a positive attitude toward smart technologies to increase digital knowledge. Moreover, the strategic design of a harmonized set of ICT-based projects engages stakeholders in the codesign of smart community.

The adoption of a sustainable and innovation-oriented entrepreneurial attitude encompasses the attainment of economic (micro), social (meso), and environmental (macro) development. The emergence of a new form of entrepreneurship based on the proposition of an open innovation business model and on the introduction of new economic and legal standards can be confirmed.

Then TrentoSmartCity governance seems to espouse a smart innovation-oriented mind-set and entrepreneurial leadership attitude settling a visionary philosophy in the entire community. The new technologies are employed to involve users, entrepreneurs, researchers, or public employees in the cocreation of value actualized in sustainable and innovative practices.

At the meso-level, smart values translate into active stakeholders' engagement in service codelivery. According to the president of Bruno Kessler foundation, the creation of new forms of social participation is a key lever to encourage innovation and social change.

TrentoSmartCity governance adopts ad hoc social participation strategies to engage citizens in decision-making nurturing the codevelopment of projects and enabling the proposition of new ideas.

The e-inclusion of citizens can strengthen the process of e-democracy addressing relevant social challenges. According to the President of Impact Hub: "Smart communities cannot exist without smart people." Digital inclusion acts as a key lever to: (1) improve digital knowledge; (2) enable the proposition of innovative insights from and for the community.

Making citizens more confident with technology can enable their active contribution to service and community improvement and can also strengthen their positive attitude toward smartness.

Users' abilities to propose innovative insights require the ongoing ideas exchange among and between citizens and digital experts. The most relevant project is "WeLive" that launches a platform subdivided into three main sections: signals, ideas, and challenges, in which users can inform the other citizens on malfunctions, can share their insights with citizens, and can propose new mobile applications based on their experience.

The role of multilevel governance in Trento smart community arises at the macro-level as a harmonizer of the relationships between and among actors. Thanks to the successful implementation of the smart projects described above, open and collaborative solutions of e-government are realized in which every actor is engaged and conceived as a peer thanks to the offering of smart technologies to make more with less and using mobile technologies to simplify the access to public services.

The creation of an e-government simplifies the relationship between citizens and local government in which users are able to identify the different stages of a public procedure and browse through the problems and solutions contributed by other users while interacting with each other.

Thus, at the macro-level, citizens do not propose "simple" suggestions and innovative insights but can perform concrete actions for community improvement (coinnovation policies). In this way, actors become policy-makers introducing new practices of policy design. Thanks to the "Welive" platform, users can also propose solutions to some "challenges" launched by TrentoSmartCity to "challenge" their abilities to address relevant societal matter.

5. Discussion

The findings obtained can lead to the proposition of a framework (see Fig. 3) that conceptualizes a metalevel that synthesizes the orchestrating role of multilevel governance and identifies the main strategic steps pursued to chase sustainable innovation at the micro-, meso-, and macro-levels to pursue economic (profit), social (people), environmental, and institutional (planet) development. These are: (1) codesign of smart culture; (2) codevelopment of smart projects and joint decision-making; (3) coevaluation of shared values and innovative ideas and suggestions from users; (4) coinnovation and proposition of sustainable-innovative policies.

The strategic codesign of smart culture (step 1) is realized thanks to the sharing of a proactive mind-set based on the diffusion of a positive attitude toward technology. Long-term strategies are conceived to engage stakeholders in the design of a harmonized set of projects in which ICTs are integrated synergistically to realize city development. The objective is to develop an integrated technological infrastructure based on an open innovation business model (economic dimension, profit) in which knowledge exchanges and relationships among actors can be managed and harmonized flexibly.

Then, based on the sharing of common smart values, users are engaged in the cocreation of decisions (step 2). The codevelopment of smart projects is realized through joint decision-making activities in which users' knowledge and

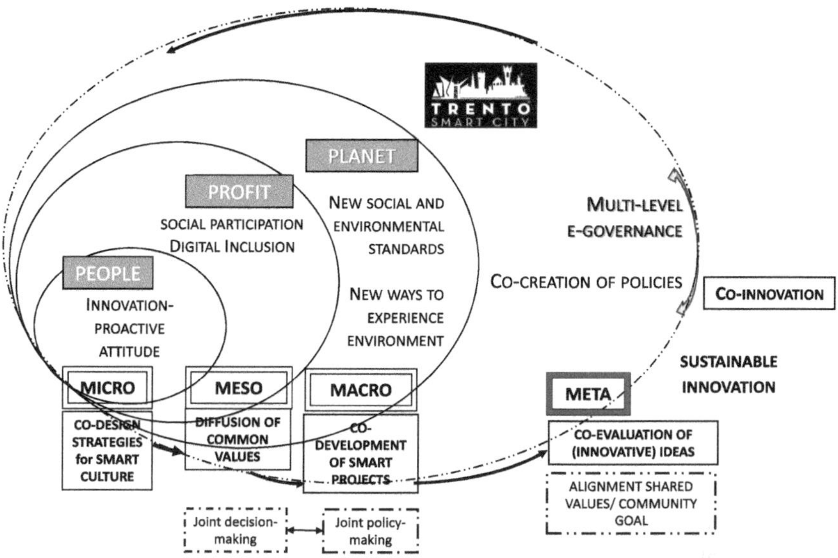

FIG. 3 The elaboration of a framework for multilevel governance in smart cities. *(Data from Author's elaboration.)*

the ideas and insights proposed are integrated thanks to trustful and transparent relationships with the other actors (social dimension, people).

The integration of innovative insights from many sources should be optimized by multilevel governance thanks to the constant search for the compliance between individual objectives and values with the overall system's goal (step 3). This alignment is the driving force that can lead to the transition from the decisions coestablished at the meso-level to the codevelopment of policies at the macro-level (institutional dimension, planet). The innovative ideas and suggestions from users should be coevaluated to align the shared values in the community with the overall goal of sustainable development.

Thus each user can become a policy-maker by turning shared decisions into common policy (step 4). The last step of the cycle is the codevelopment of policies and of innovative solutions to address opportunities, innovation and to strengthen the culture of the city. Thanks to democratic governance, new open models based on innovation and e-governance can be established and new cultural and social institutions can be created.

Lastly, the process of coinnovation as a cycle can be renewed over time (the arrow in the figure). The participatory and multilevel governance of TrentoSmartCity can regenerate the smart culture produced through the constant renewal of values and knowledge thanks to the adaptation to environmental changes and the codesign of common foundations for a smart future.

6. Theoretical and managerial implications

In line with a systems and process-based view, the study shows that the main steps accomplished by multilevel governance in smart cities can start from the diffusion and codevelopment of a cohesive culture to advance and implement policies according to a participatory approach. Definitively, the research shows that governance's key role should start from the establishment of a set of common values at the micro-level and that ICT's efficient use can benefit from an innovation-oriented mind-set to create multiple advantages throughout the city at an economic, relational, educational, and societal level.

Therefore, from a theoretical standpoint, the investigation of the main ecosystem's elements that act in Trento Smart community can guide researchers in the identification of the main drivers for growth and sustainable innovation (Kourtit et al., 2012) in smart cities. In addition, the identification of a human component as a key enabler of sustainable development is in line with a recent trend in the literature that considers technology as a means for sustainability and innovation and not as the end of the process (Visvizi et al., 2018; Visvizi & Lytras, 2018).

From a managerial standpoint, the framework guides management and governance in the elaboration of strategies to employ smart technologies wisely and to exploit the different possibilities offered by the diverse kinds of existing technological tools (cognitive computing to track or predict users' perception, recommendation systems to store users' behavior, or blockchain technologies to foster service improvement). By associating each ecosystem's context with the detection of different sustainable innovation practices, smart cities managers can understand how to elaborate targeted strategies to enhance cocreation, strengthen relationships between actors, and increase well-being in each level and for each engaged stakeholder's group.

The study shows some limitations related to the adoption of case study methodology that does not allow any generalization of results. Thus the findings obtained cannot be extended to other contexts. However, by adopting an exploratory approach, the research performed in this work can be considered as a first step that can address future research toward the application of the general framework proposed to other smart cities or to other smart sectors (mobility, environment, energy, etc.) through comparative case studies.

7. Conclusions

The integrated framework obtained from the re-elaboration of results can contribute to assess the main steps, objectives, and implications related to the strategic use of smart technologies in cities by revealing their impact on social, economic, and environmental spheres.

Empirical research investigates the different sustainable outcomes developed in TrentoSmartCity in the different ecosystem contexts (micro, meso, macro) and adopts a synthesis view for the exploration of governance strategies to foster sustainable innovation.

Firstly, for each ecosystem context and sustainability dimension (economic, social, political-environmental), the various actors, technologies, resources, institutions, and value proposition are explored. This multileveled analysis contributes to detecting the different actors' practices that can enhance value cocreation and, thus, innovation and sustainable innovation at the Triple Bottom Line (profit, people, planet). This can lead to the introduction of a general framework which rereads smart contemporary cities through the definition of the technology-mediated interactions that foster innovation to compare the different practices occurring in different systems (private companies, public organizations, nonprofit association, research system, etc.).

Then a synthesis perspective is employed to examine how governance can manage strategically the resource integration and to optimize the creation of new value and knowledge across the three contexts and sustainable dimensions.

The multilevel governance implemented in TrentoSmartCity can regenerate the smart culture produced in the ecosystem through the constant renewal of values and knowledge thanks to the adaptation to social challenges, environmental and institutional requirements, and the codesign of common foundations for a smart future. Lastly, TrentoSmartCity owns the ability to address social challenges and create opportunities from them, by making the city resilient and sustainable. Then, by meeting the aim established by the 11th Sustainable Development Goal, the case study of Trento shows that sustainable innovation is the key lever to guarantee smart cities growth and development.

References

Akaka, M. A., Vargo, S. L., & Lusch, R. F. (2013). The complexity of context: A service ecosystems approach for international marketing. *Journal of International Marketing*, *21*(4), 1–20.

Albino, V., Berardi, U., & Dangelico, R. M. (2015). Smart cities: Definitions, dimensions, performance, and initiatives. *Journal of Urban Technology*, *22*(1), 3–21.

Appio, F. P., Lima, M., & Paroutis, S. (2019). Understanding smart cities: Innovation ecosystems, technological advancements, and societal challenges. *Technological Forecasting and Social Change*, *142*, 1–14.

Barile, S., Ciasullo, M. V., Troisi, O., & Sarno, D. (2017). The role of technology and institutions in tourism service ecosystems: Findings from a case study. *The TQM Journal*, *29*(6), 811–833.

Barile, S., Saviano, M., Polese, F., & Di Nauta, P. (2013). Il rapporto impresa-territorio tra efficienza locale, efficacia di contesto e sostenibilità ambientale. *Sinergie Italian Journal of Management*, *90*, 25–49.

Bătăgan, L. (2011). Smart cities and sustainability models. *Informatica Economică*, *15*(3), 80–87.

Caragliu, A., Del Bo, C., & Nijkamp, P. (2011). Smart cities in Europe. *Journal of Urban Technology*, *18*(2), 65–82.

Chandler, J. D., & Vargo, S. L. (2011). Contextualization and value-in-context: How context frames exchange. *Marketing Theory*, *11*(1), 35–49.

Dulock, H. L., & Holzemer, W. L. (1991). Substruction: Improving the linkage from theory to method. *Nursing Science Quarterly*, *4*(2), 83–87.

Eisenhardt, K. M. (1989). Building theories from case study research. *Academy of Management Review*, *14*(4), 532–550.

Elkington, J. (1998). Partnerships from cannibals with forks: The triple bottom line of 21st-century business. *Environmental Quality Management*, *8*(1), 37–51.

Giffinger, R., Fertner, C., Kramar, H., Meijers, E., & Pichler Milanovic, N. (2007). *Smart cities: Ranking of European medium-sized cities*. Vienna: University of Technology.

Gil-Garcia, J. R., Zhang, J., & Puron-Cid, G. (2016). Conceptualizing smartness in government: An integrative and multi-dimensional view. *Government Information Quarterly*, *33*(3), 524–534.

Harrison, C., Eckman, B., Hamilton, R., Hartswick, P., Kalagnanam, J., Paraszczak, J., et al. (2010). Foundations for smarter cities. *IBM Journal of Research and Development*, *54*(4), 1–16.

Höjer, M., & Wangel, J. (2015). Smart sustainable cities: Definition and challenges. In L. M. Hilty, & B. Aebischer (Eds.), *ICT innovations for sustainability* (pp. 333–349). Cham: Springer.

Kitchin, R. (2015). Making sense of smart cities: Addressing present shortcomings. *Cambridge Journal of Regions, Economy and Society*, *8*(1), 131–136.

Komninos, N. (2013). *Intelligent cities: Innovation, knowledge systems and digital spaces*. Abingdon-on-Thames, UK: Routledge.

Kooiman, J., Bavinck, M., Chuenpagdee, R., Mahon, R., & Pullin, R. (2008). Interactive governance and governability: An introduction. In J. Edelenbos, & I. Van Meerkerk (Eds.), *Critical reflections on interactive governance. Self-organization and participation in public governance* (pp. 29–50). Cotswolds: Edward Elgar Publishing.

Kourtit, K., Nijkamp, P., & Arribas-Bel, D. (2012). Smart cities perspective—A comparative European study by means of self-organizing maps. *Innovations*, *25*(2), 229–246.

Krippendorff, K. (2004). Reliability in content analysis: Some common misconceptions and recommendations. *Human Communication Research*, *30*(3), 411–433.

Lim, C., Kim, K. J., & Maglio, P. P. (2018). Smart cities with big data: Reference models, challenges, and considerations. *Cities*, *82*, 86–99.

Lombardi, P., Giordano, S., Farouh, H., & Yousef, W. (2012). Modelling the smart city performance. *Innovation: The European Journal of Social Science Research*, *25*(2), 137–149.

Lytras, M. D., Raghavan, V., & Damiani, E. (2017). Big data and data analytics research: From metaphors to value space for collective wisdom in human decision making and smart machines. *International Journal on Semantic Web and Information Systems (IJSWIS)*, *13*(1), 1–10.

Lytras, M., & Visvizi, A. (2018). Who uses smart city services and what to make of it: Toward interdisciplinary smart cities research. *Sustainability*, *10*(6), 1998.

Malek, J. A. (2009). Informative global community development index of informative smart city. In *Proceedings of the 8th WSEAS international conference on education and educational technology* (pp. 17–19). Athens, Greece: WSEAS Press.

Marsal-Llacuna, M. L., Colomer-Llinàs, J., & Meléndez-Frigola, J. (2015). Lessons in urban monitoring taken from sustainable and livable cities to better address the smart cities initiative. *Technological Forecasting and Social Change*, *90*, 611–622.

Mellouli, S., Luna-Reyes, L. F., & Zhang, J. (2014). Smart government, citizen participation and open data. *Information Polity*, *19*(1,2), 1–4.

Nam, T., & Pardo, T. A. (2011). Conceptualizing smart city with dimensions of technology, people, and institutions. In *Proceedings of the 12th annual international digital government research conference: Digital government innovation in challenging times* (pp. 282–291). New York, United States: Association for Computing Machinery (ACM).

Palmisano, S. J. (2008). *A smarter planet: The next leadership agenda. Vol. 6* (pp. 1–8). Istanbul, Turkey: IBM. November.

Paroutis, S., Bennett, M., & Heracleous, L. (2014). A strategic view on smart city technology: The case of IBM smarter cities during a recession. *Technological Forecasting and Social Change*, *89*, 262–272.

Pereira, V. G., Cunha, M. A., Lampoltshammer, T. J., Parycek, P., & Testa, M. G. (2017). Increasing collaboration and participation in smart city governance: A cross-case analysis of smart city initiatives. *Information Technology for Development*, *23*(3), 526–553.

Piciocchi, P., Spohrer, J., Bassano, C., & Giuiusa, A. (2012). Smart governance to mediate human expectations and systems context interactions. In J. C. Spohrer, & L. E. Freund (Eds.), *Advances in the human side of service engineering* (pp. 319–328). Oxfordshire, UK: Taylor & Francis.

Polese, F., Troisi, O., Carrubbo, L., & Grimaldi, M. (2018). An integrated framework toward public system governance: Insights from viable systems approach. In A. Bonomi Savignon, L. Gnan, A. Hinna, & F. Monteduro (Eds.), *Cross-sectoral relations in the delivery of public services* (pp. 23–51). West Yorskshire, UK: Emerald.

Spohrer, J., Piciocchi, P., & Bassano, C. (2012). Three frameworks for service research: Exploring multilevel governance in nested, networked systems. *Service Science*, *4*(2), 147–160.

Stoker, G. (2006). Public value management: A new narrative for networked governance. *American Review of Public Administration*, *36*(1), 41–57.

Storbacka, K., Brodie, R. J., Böhmann, T., Maglio, P. P., & Nenonen, S. (2016). Actor engagement as a microfoundation for value co-creation. *Journal of Business Research*, *69*(8), 3008–3017.

Stratigea, A. (2012). The concept of 'smart cities'. Towards community development? *Netcom Réseaux, Communication et Territoires*, *26*(3/4), 375–388.

Tellis, W. M. (1997). Application of a case study methodology. *The Qualitative Report*, *3*(3), 1–19.

Toppeta, D. (2010). The smart city vision: How innovation and ICT can build smart, "livable", sustainable cities. *The Innovation Knowledge Foundation*, *5*, 1–9.

Vanolo, A. (2016). Is there anybody out there? The place and role of citizens in tomorrow's smart cities. *Futures*, *82*, 26–36.

Vargo, S. L., & Lusch, R. F. (2011). It's all B2B… and beyond: Toward a systems perspective of the market. *Industrial Marketing Management*, *40*(2), 181–187.

Vargo, S. L., & Lusch, R. F. (2016). Institutions and axioms: An extension and update of service-dominant logic. *Journal of the Academy of Marketing Science*, *44*(1), 5–23.

Visvizi, A., & Lytras, M. (2018). Rescaling and refocusing smart cities research: From mega cities to smart villages. *Journal of Science and Technology Policy Management*, *9*(2), 134–145.

Visvizi, A., & Lytras, M. (Eds.). (2019). *Smart cities: Issues and challenges: Mapping political, social and economic risks and threats*. Amsterdam: Elsevier.

Visvizi, A., Lytras, M., Damiani, E., & Mathkour, H. (2018). Policy making for smart cities: Innovation and social inclusive economic growth for sustainability. *Journal of Science and Technology Policy Management*, *9*(2), 126–133.

Yin, R. (1994). *Case study research*. California: Sage Publications.

Chapter 7

Open data portals for urban sustainable development: People, policies, and technology

Christine Meschede[a] and Tobias Siebenlist[b]
[a]*Department of Information Science, Heinrich Heine University Düsseldorf, Düsseldorf, Germany,*
[b]*Ronin Institute for Independent Scholarship, Montclair, NJ, United States*

1. Introduction: Open data, cities, and the SDGs

With the Sustainable Development Goals (SDGs), the United Nations introduced 17 global goals to end poverty, protect the planet, and ensure prosperity for all (UN, 2015b). These goals, adopted in 2015, are intended to unite various dimensions of sustainable development—including social, environmental, and economic aspects—in one agenda. Each of the 17 goals consists of further concrete objectives, adding up to 169 targets and more than 200 indicators for measuring progress thereon. In contrast to the Millennium Development Goals (MDGs), preceding the SDGs, these SGDs demand action on the part of all countries, every country, regardless of its stage of development status. Apart from that, the SDGs, for the first time, directly embed the importance of cities for sustainable development. With SDG 11 (sustainable cities and communities), the goals draw attention to the global trend toward urbanization. This "Urban Sustainable Development Goal (USDG)" (Klopp & Petretta, 2017) is further linked to all other SDGs and their targets. The New Urban Agenda, adopted in 2016 during the UN Conference on Housing (Habitat III), provides guidelines for future urban development and also highlights the role of cities for sustainable development by referring to SDG 11 (UN, 2016).

Governance for sustainable development has grown immensely during the past years, and several national sustainable development strategies have been adopted (Quental, Lourenço, & Da Silva, 2011). Similarly, municipalities adopt local sustainable development strategies and thus support the SDGs at the local level. Some cities count on citizen participation for developing and implementing such a policy and further stress the importance of open data in this context (Meschede & Mainka, 2020). Open data, i.e., data that is machine-readable, freely available, and reusable for everyone, is intended to increase transparency

Smart Cities and the UN SDGs. https://doi.org/10.1016/B978-0-323-85151-0.00007-5

and trust toward governments. SDG 16 (peace, justice, and strong institutions) emphasizes the role of information accessible for everyone with target 16.10: "Ensure public access to information and protect fundamental freedoms, in accordance with national legislation and international agreements" (UN, 2015a). In the context of smart cities, information, and, in particular, information and communication technology (ICT) play a major role. By connecting the targets of SDG 11 and SDG 16, the significance of cities striving toward becoming smart sustainable cities comes to the fore. Moreover, open data can also serve as an instrument for measuring progress toward the SDGs and their targets. Several initiatives and organizations report on the achievement of the SDGs on a global scale. For example, the SDG National Reporting Initiative (2018) aims at facilitating information sharing on the SDGs among policymakers, data managers, and international organizations. The Atlas of Sustainable Development Goals, a report, and visualization on the SDGs' targets is a project of the World Bank and provides global development data (WB, 2018). Open Data Watch, a nonprofit, nongovernmental organization, provides the Open Data Inventory (ODIN) assessing the coverage and openness of official statistics regarding social, economic, and environmental information and ranks several countries in this regard (Open Data Watch, 2017). ODIN thus explicitly analyses data categories oriented toward progress on the SDGs but does not concentrate on the concrete targets of the goals. In 2018, the highest overall score was achieved by Singapore, whereas Denmark and Sweden occupied the first rank in 2017 and 2016, respectively. All results and datasets are available under the Creative Commons license (ODIN, 2020).

At the city level, an ongoing challenge related to the implementation of the SDGs is the limited availability of urban data (Klopp & Petretta, 2017). By opening more and more datasets, open urban data can be appropriate to close this gap (Meschede & Siebenlist, 2019). Therefore, a framework is needed to map existing datasets to the targets and indicators of the SDGs. The Interagency and Expert Group on SDG Indicators (IAEG-SDGs) has developed the underlying indicator framework. The indicator set is not a fixed system; it is reviewed and updated regularly. This fact poses challenges for measuring progress toward the SDGs and providing sufficient data. Solutions are needed to display information and data for the SDGs adequately and to make it comparable, especially at the city level. There exist some open data portals that are designed for this purpose. In the next section, we illustrate the background of open data portals from different perspectives.

2. Open urban data from different perspectives

By providing data on sustainable development that is available and reusable for everyone, as well as machine-readable, citizens (people) are sensitized toward this topic and may reuse the datasets for useful applications and visualizations. Governments may push this development and encourage the opening of datasets (policies). Therefore, technical solutions in the form of adequate open data portals are necessary (technology). In the following, we describe the prerequisites of open data on sustainable development from the perspectives of people, policies, and technology.

2.1 People

Creating added value for citizens is among the core targets of open data (Attard, Orlandi, Scerri, & Auer, 2015). The concept of open government, which entails efforts toward open data, mainly concentrates on two dimensions, which are claimed to be beneficial for all people, i.e., transparency and participation (Meijer, Curtin, Hillebrandt, & Curtin, 2012). In contrast, smart city research has, to some extent, been criticized for mainly concentrating on technology and ignoring the citizen-perspective. Although several smart city services are widely used, there are still technical, behavioral, and policy-related barriers that are affecting potential users. In particular, security and data privacy are essential factors for the adoption of smart city services (Lytras & Visvizi, 2018). Still, data-driven approaches are significant aspects of smart cities and can contribute to sustainable urban development (Francisco & Taylor, 2019). By combining the concept of open data with data-driven strategies concerning sustainability issues, the citizen-perspective in smart cities can be strengthened, thus emphasizing the importance of people living in a city. Open data initiatives can lead to an enhanced public value in smart cities, e.g., by controlling diseases and pandemics or by increasing the quality of public services (Pereira, Macadar, Luciano, & Testa, 2016). Open data platforms for health parameters can additionally improve living in smart cities. Smart health approaches, including monitoring and visualizing health-related data, and providing information through stored health data, aim to promote education and awareness concerning citizens' health and lifestyle (Trencher & Karvonen, 2019). Another example demonstrates that open urban energy data can improve the understanding of energy use and increase transparency and awareness concerning this topic (Francisco & Taylor, 2019). Citizens can further become "data pro-sumers (both consumers and providers of data)" (Charalabidis, Loukis, & Alexopoulos, 2014, p. 2114). For example, participatory approaches to measuring fine particulate matter can enhance the quality of living in smart cities (Chen et al., 2017). For encouraging citizens to participate in processes like this and to interact with open data, solutions that provide residents and public administration common ground for discussions are essential (Scarano et al., 2015). Governments or administrations can directly initiate participation opportunities for citizens, and thus co-creation processes can be initiated. Alternatively, interested citizens can become involved in civil society organizations that deal with such topics as open government and open data. An example is the Open Knowledge Foundation (Open Knowledge Foundation, 2020), which offers broad opportunities for participation and is represented by international chapters and established groups in many countries.

The context and people involved are essential factors for discussing the meaning of open data. However, there seem to be inconsistencies between the offered types of data and the actual expectations and needs of users. Simply opening datasets that nobody is interested in is therefore not enough and might only lead to "nominal transparency" (Ruijer & Grimmelikhuijsen, 2020). Data concerning sustainable urban development and the SDGs is of interest to every citizen, as this kind of data reports on the quality of urban life across several dimensions that affect everyone living in a particular municipality. Besides, re-using such data for useful applications and services can further contribute to an improved quality of life for the citizens. As a result, adequate ways of creating such added values for citizens are necessary. In the context of smart cities, "the city of the knowledge society can go a step further, by adapting itself to citizens instead of citizens having to adapt to the city" (Pérez del Hoyo & Mora, 2019, p. 188). National, regional, and local policies can contribute to developing the framework conditions for this adaption process including open data, transparency, and participation. Finally, these policies have to be translated into concrete, practical solutions. The following two subsections discuss these two dimensions.

2.2 Policies

The publication of data as open data is not an end in itself and will not be undertaken by interested countries or cities as a stand-alone measure. The path toward more openness and transparency of governments (Open Government) is accompanied by the publication of data, among other things. According to Nugroho, Zuiderwijk, Janssen, & de Jong, 2015, "Open data policies guide the opening and stimulate the usage of public data" (A comparison of national open data policies: lessons learned). Three waves are mentioned: opening data, using data, and developing added value based on the data. Policies are needed to advance efforts toward the publication and provision of open data through portals with a shared understanding within each country. Zuiderwijk and Janssen (2014) have developed a framework for the comparison of open data policies at different levels of government, which takes different aspects into account and thus enables a comparison of different policies and serves as well a starting point for the development of new policies. This chapter does not have an exclusive focus on the comparison of open data policies; therefore, the framework is not applied. Instead, we use data already collected from the Open Data Barometer (2020).

In the current version of its report (the Leaders edition) the Open Data Barometer concludes: "Government strategy and policies: Open data strategies and policies are essential to articulate processes and responsibilities, and yet, this edition shows that governments typically build these only after open data initiatives are already in place" (WWWF, 2018). It can thus be seen that an opening of data has often already taken place in various areas before official regulations are introduced. This does not need to harm the process, it can be accelerated, and initial experiences can be drawn on. However, there is a risk that without a common strategy, individual participants may take many different paths and establish meaningful discrepancies in terms of, for example, which data is published in which formats and under which licenses.

The Sunlight Foundation has created a living set of open data guidelines that provide a good first insight. These point out a possible direction and a good example for the publication of data. It is divided into three major sections: "What data should be public," "How to make data public," and "How to implement policy." Here it is the last section, in particular, that is to be considered. It is not only about prescribing the publication of data, but also about regular quality checks and the inclusion of citizens' opinions. The long-term financing of such projects should also be regulated, not least, to ensure sustainable operation and further development. The establishment of partnerships with different stakeholders, such as "… nonprofits think tanks, academic institutions, and nearby governments …" (Sunlight Foundation, 2014) can be beneficial to investigate and continuously improve the efforts. It also helps to establish contact with potential users, so that the creation of added value can be addressed at this level. If civil society organizations such as the Open Knowledge Foundation are already involved here, direct contact with citizens can be established so that the aspects of policies and people come together here, and participation is made possible. On this basis, open innovation projects can take place (Hartmann, Mainka, & Stock, 2019).

Many countries that are seriously committed to the principles of open government and are working on its implementation have now also joined the Open Government Partnership (2020). The Open Government Partnership is a partnership to promote accountable, responsive, and inclusive governance founded in 2011. The partnership consists of 78 countries and 20 local members that work alongside thousands of civil society organizations. Every 2 years, the member countries adopt national action plans in which targets for improving open government efforts are set. Open data also plays an essential role in this open government process, which is developed via co-creation in cooperation with civil society organizations. Based on these plans and the achievement of the goals that they have set themselves, the path to a better open government implementation can be presented in a transparent and comprehensible manner. This represents an extension of existing policies and can be adapted in a 2-year cycle depending on progress. Such an approach enables joint learning and promotes a common understanding, provided that the co-creation and the implementation phases are designed in a cooperative and participatory manner.

Across the world, the degree of implementation of open government ideas and open data portals varies greatly. However, besides many regional and local conditions that are responsible for the differences, existing open data portals and policies can be used to establish rankings for the maturity of open data. One example is the Open Data Barometer. In its last edition (the leaders' edition) from September 2018, the report looks specifically at 30 governments that have made concrete commitments to champion open data, either by adopting the Open Data Charter, or, as members of the G20, by signing up to the G20 Anti-Corruption Open Data Principles (https://opendatabarometer.org/leadersedition/report/). In contrast to the fourth edition of the report, this report explicitly looks at governments that have already made or are making significant progress and can, therefore, be considered good examples of the introduction and implementation of policies. The results can be found in the Open Data Barometer. The "readiness" index includes government policies, which account for one-quarter of the readiness index. Therefore, this value is of particular interest in this section.

The Open Data Barometer report ends with a short outlook into the future, in which policy recommendations are given. Here, three central points are mentioned whose implementation is considered particularly important: (i) put "open by default" into action, (ii) build and consolidate data infrastructure, and (iii) publishing with purpose. The first point deals with further development from pure policies toward guidelines and action items. Process, responsibilities, and timelines should be defined, and the necessary financial and human resources planned and made available. "Open by default" is described as an aspired goal that cannot be achieved overnight. However, it helps to keep this in mind in order to lead the development step by step in a direction where collected data directly becomes open data. The second point addresses both the skills of employees and the adaptation of technical systems. Employees in administrations must have the necessary skills to generate open data and to handle them with confidence. To this end, training of existing employees, the recruitment of suitable staff and the right mindset must be promoted. On the technical side, software portals for providing open data must be further developed and kept as interoperable as possible. The systems from which the data of administrations and governments originate do not usually support this process. Such systems must be adapted accordingly, i.e., equipped with interfaces and processing options or replaced by suitable programs that support the process. For postpublication systems, quality standards must be defined and adhered to, and possibilities must be created to control data-driven processes via APIs. These changes are also not possible all at once, and often longer transformation processes are necessary. However, it is always helpful to keep the goal in mind to take these criteria into account when tendering for and purchasing new programs and systems, thus gradually improving the situation. The third point relates directly to interaction with people. It is not merely a matter of publishing any data. Of course, as much data as possible should be made available as open data. However, when prioritizing and sequencing, the question arises as to the purpose for which this data can be used. For this purpose, contact with future users—civil society organizations, interested citizens, companies—is critical to understand the needs and wishes. This understanding is not necessarily available a priori but must also be promoted by the administrations. In the case of sustainable urban development, data on the SDGs at the local level is particularly necessary. In addition to the pure—relatively simple—provision of data on portals, it is also a matter of identifying and defining challenges and opportunities. The questions of "why" and "what for" come to the fore and must be answered meaningfully by both sides, so that published data and needs or ideas can ultimately result in added value.

At these three points, the three aspects' coherence in this chapter becomes very clear in this chapter. Policies are necessary to control and drive the publication of data. This is achieved in cooperation processes with people who are enabled to experience transparency and participation. The data and portals are digital and from them—mostly but not exclusively—digital products or services are created, which both require technical support and are, in turn, used by people.

Following on from the first point in the policy recommendations of the Open Data Barometer, other areas that can be seen in the context of smart sustainable cities also play an important role in addition to the publication and processing of data in the area of Open Government. For example, there is an initiative "Public Money Public Code" of the Free Software Foundation Europe (https://fsfe.org/), which demands that software developed and paid for by public money must be made available as open-source. The demand can be briefly summarized as follows: "We want legislation requiring that publicly financed software developed for the public sector be made publicly available under a Free and Open Source Software license" (https://publiccode.eu). Again, the same values apply here as before:

- Transparency for administrations as well as for interested citizens regarding the functionality of the software.
- Participation: Further development and adaptation can be taken over by oneself or passed on to other competent persons or companies, vendor lock-in is avoided.
- Sustainability: Software/code can be passed on, other cities with the same needs do not have to develop new software and code can be optimized for energy efficiency.
- Data sovereignty: When software projects are awarded, the data collected is often also passed on to the provider, and publication is made more difficult; by using open source software with appropriate licenses and agreements, data sovereignty remains with the cities concerned, making it easier to publish and reuse data.

The city of Barcelona is the first city council who signed the open letter for this initiative and can be considered a pioneer in this field (https://fsfe.org/news/2018/news-20180705-01.de.html).

2.3 Technology

In addition to the real advancement of the opening and provision of data, the actual implementation plays an enormously important role when it comes to enabling subsequent use as conveniently and freely as possible. This, in turn, applies to both sides of the use, both for the providers and users. Various portal solutions for the provision of open data have emerged—some as open-source software and others as commercial products. In contrast to the provision of data on a simple website for download, specialized portals offer additional possibilities such as search and filter options, the inclusion of metadata, classification into categories, and preview views of the data contained. The published data should comply with the open definition, i.e., be machine-readable, available in an open format, and reusable. When using different portal software, care should be taken to ensure that interoperability is at least possible or can be made possible. In concrete terms, this means that appropriate import and export interfaces are available and that operation via an API is possible.

Concerning licenses, it is advisable to think about their use in advance. When choosing the license, it is crucial to ensure that it is a free license that does not restrict the use of the data. Otherwise, the data is not open data in the real sense of the word, and the step toward creating added value is blocked. There are already several internationally used licenses (e.g., Creative Commons) that meet these criteria. To avoid a proliferation of different and possibly incompatible licenses, it would be more than desirable to follow international standards. Otherwise, many national licenses will be created, for which the compatibility with other licenses must be checked in each case and unnecessarily complicate or prevent use. The same applies to the reproduction of, for example, the CC licenses with an own, very similar licenses. The latter is unnecessary if it covers the same purposes.

Open data is published and made available at different administrative levels depending on the country. This is briefly shown using the example of Germany as a federal country. There are roughly three levels at which data are published: cities/municipalities, federal states, and state. Federalism can be seen as a strength or weakness, as it brings its developments and challenges. However, this chapter does not discuss the federalist principle. An open data law currently exists only at the federal level. The individual federal states are progressing at different speeds, and the same applies to individual municipalities and cities. It often depends on the individual actions and decision-makers how the topics open government and open data develop. Thus, there are quite small cities that can be described as pioneers in the field of open data and big cities that are still at the beginning of these developments.

In any case, it is helpful in the strategy for publishing data if universal principles, procedures, and standards can be agreed upon. One approach to this is through an initiative of organizations, open data teams of some cities with the participation of civil society. These have created the first version of a model data catalog for the federal state of North Rhine-Westphalia, Germany, which can serve as a guide to what data is already commonly published (https://musterdatenkatalog.de). This can be particularly useful when starting new open data initiatives but also generally helpful in understanding and proceeding around the opening of data.

If this idea is taken further, thematic catalogs are the logical next step, in addition to a general sample data catalog. For example, separate sample data catalogs could be created for the individual categories in open data portals or special topics such as the Sustainable Development Goals. This would then be the use of finding the mapping of the goals as an independent thematic field or subcategory found through tags in a general data portal. On this basis, the necessary data for calculating the achievement of the goals at the city level can be published, found, and finally used.

To conclude, opening urban datasets on sustainable development for the public requires adequate platforms and prepared high-quality datasets. Without sufficient datasets, collaborative learning processes can be discouraged (Ruijer & Grimmelikhuijsen, 2020). There are two main possibilities for providing open data concerning the SDGs at the municipal level: (1) the integration of SDG related datasets and indicators in existing open data portals or (2) the development of specific sustainable development portals solely addressing the SDGs. Currently, both approaches are observable in different countries that are providing sustainable development data at the national level. In the following, we will present various examples at the national level and discuss the pros and cons of each. We selected SDG data portals from three countries out of each category of the Open Data Barometer (Champions, Contenders, Stragglers), exploring the differences between them. Table 1 summarizes the main characteristics of these countries' open data portals. The applicability at the city level of the approaches depicted below will be discussed in the subsequent section.

2.3.1 Open data champion: Australia

Australia ranks third according to the Open Data Barometer and can be viewed as a champion in handling open data. According to ODIN, Australia achieves a score of 68 (out of 100) concerning the coverage and openness of data on sustainable development. The Australian government provides a general open data platform (https://data.gov.au/data). Searching for the SDGs, one can find three datasets tagged with this term (May 16, 2020). All three datasets are related to violence and emotional and sexual abuse, with two datasets concentrating on women and thus addressing SDG 5 (gender equality). The Australian government does not only provide these three datasets concerning the goals but also hosts a specific platform on the SDG indicators (https://www.sdgdata.gov.au/). For each goal, the country reports on the specific indicators, if they are among the globally, agreed-upon relevant targets. The data portal classifies for each indicator if (1) the corresponding dataset is available, (2) work is underway to publish the dataset, (3) no such dataset is available, or if (4) it is not applicable due to the globally agreed methodology. Taking a deeper look at the urban sustainable development goal, SDG 11, out of 15 indicators, six are marked as not applicable. The remaining nine indicators are classified into category 2—work is underway to publish the corresponding dataset (May 16, 2020).

TABLE 1 Comparison of the open data portals of Australia, Germany, and South Africa.

	Australia	Germany	South Africa
General Open Data Portal	https://search.data.gov.au	https://www.govdata.de/	https://southafrica.opendataforafrica.org/
SDG Data Platform	https://www.sdgdata.gov.au	https://sustainabledevelopment-germany.github.io/en/	https://southafrica.opendataforafrica.org/vaaasvg
Responsibility for SDG Data Platform	Australian Government: Department of Foreign Affairs and Trends	German Federal Statistical Office	Republic of South Africa, Department of Statistics African Development Bank Group
Availability of datasets concerning SDG 11 (May 16, 2020)	0	8	0
ODIN score (2018) (max. 100) • Social statistics (max. 100) • Economic statistics (max. 100) • Environmental statistics (max. 100)	68 • 62 • 79 • 65	79 • 68 • 89 • 81	51 • 36 • 58 • 61
Category Open Data Barometer (2018) • Total score (max. 100) • Readiness (max. 100) • Implementation (max. 100) • Impact (max. 100)	Champion • 75 • 70 • 84 • 62	Contender • 58 • 76 • 72 • 27	Struggler • 36 • 50 • 37 • 22

2.3.2 Open data contender: Germany

Germany is in the midfield concerning open data maturity and thus rated as a contender by the Open Data Barometer. With an overall score of 79 (out of 100), Germany scores relatively well regarding the availability and openness of statistical data on sustainable development. The general data portal on open government data in Germany is hosted by GovData, located at the fiscal authority of the city of Hamburg (https://www.govdata.de/). Searching for the Sustainable Development Goals does only provide three pdf files, all representing sustainable development reports of the Federal Ministry of Defense and the German Armed Forces. Concrete datasets providing data on the SDGs' indicators are not available. Similar to Australia, Germany also hosts a specific portal solely addressing the goals. The German SDG Data Platform is provided by the German Federal Statistical Office (https://sustainabledevelopment-germany.github.io/en/). Similar to the Australian SDG platform, four categories can be divided showing the progress of the availability of datasets for each indicator. Concerning SDG 11, eight datasets are currently reported online (category 1), one dataset will be published soon (category 2), six datasets are not yet available (category 3), and none of the indicators is recognized as not applicable (May 16, 2020).

2.3.3 Open data struggler: South Africa

With an overall score of 36 (out of 100), South Africa is in the group of strugglers in the Leaders Report. It should be noted that the readiness score of 50 is substantially higher. The project Open Data for Africa is sponsored by the African Development Bank Group. The data portal for South Africa and can be reached at https://southafrica.opendataforafrica.org/. In contrast to Australia and Germany, there is a direct section on the SDGs, under which the indicators can be found, but only a few datasets are available. For SDG 11, no dataset exists.

3. Discussion and conclusion: Smart sustainable cities and open data

Cities striving to become smart sustainable cities must combine the achievements of smart cities and of sustainability or sustainable cities in a synergetic combination that not only takes both areas into account but is more than the sum of the two parts.

Both—smart concepts and sustainability—require (digital) information and data to create transparency, enable participation, and create added value. Much of this happens on a technological basis, the corresponding infrastructure must be given and freely usable. The local measurement of the achievement of the SDGs' targets gives citizens an insight into their local circumstances concerning sustainable development. They can inform themselves and demand specific improvements from their governments or initiate projects themselves that support the improvement of goal attainment. Local data and local projects are better suited for this than measurements at the national level. Thus, good ideas and projects often arise from the actions of individual or less active actors. These projects can then serve as best practices for other cities. Changes in one's environment are much more immediate and tangible than at the national level. Awareness of sustainability and the desire for reasonable technological solutions in the context of open government, in dealing with services, and companies in one's city becomes stronger as people become increasingly digitized.

The prerequisite for actions that contribute to measuring improvement is data. In most cases, this data has already been collected, but it must be published as open data. As described above, there are two ways of doing this: Integration of the allocation and information on the SDGs into existing open data portals or the creation of portals that exclusively hold data on the SDGs. In either case, there should be a leading source for the data to ensure accountability and avoid unnecessary duplication. How the path to the data of the SDGs is then taken is another question. This means, for example, that all the necessary data is available in the central, already existing open data portal, but that the labeling and search options for special categories may not be available there. In this case, a specialized portal can be useful, which provides information about the SDGs, searches, and sorts the data according to the goals, targets, and indicators, but obtains the actual data from the central open data portal. There are no copies of the files available, but instead, a link or API call is made directly to the one leading portal. If additional filters and search options are integrated into the central portal, the options may become confusing at some point. However, with a little more focus on appealing user experience, this could probably be kept within limits. Moreover, such additional options could make more people interested in topics that they have not encountered before. In this case, tags could be used in existing portals, which could be used as keywords for the appropriate data records and make them findable. This would then lead to a further filter option. Which of the two options is better accepted by users would need to be investigated in further studies.

As a starting point, urban sustainable development requires data to be able to determine transparently the situation concerning various aspects of one's city. This data is the key to transparency but also enhanced participation. However, this is only true if the data is open and can be freely used to obtain information and, based on this, create added value for specific purposes. In this chapter, the focus is set on sustainable urban development, expressed in particular through the Sustainable Development Goals of the United Nations. The need to measure and improve the SDGs at the local level is obvious: citizens can perceive and change their environment much more directly and in a more tangible way than affecting the situation for the whole country. Besides, the values at the national level only ever represent an average that does not take into account particularly progressive cities (and thus particularly lagging cities). However, these findings can play an essential role in the context of urban development processes in which the principles of open government can be taken into account purely from the development aspect. As a result, urban communities can become active civil societies in which citizens are actively involved and thus have more interest and trust in government and administrations. However, they must be empowered to do so. Open data's sustainable availability is a vital step along this path, and the consideration of specific issues such as SDGs is one element. The three issues addressed in this chapter—people, policies, and technology—are inextricably linked and interdependent and cannot be considered in isolation.

When comparing different countries, according to the criteria, it can be noted that different rankings have different criteria and thus the results come in a different order. The values from the Open Data Barometer used here are, therefore, only one possible example, which, however, also covers the policy aspect. The portals in the individual countries handle the SDGs in different ways: some integrate the data or information on the targets while others have separate portals. The city destinations are perceived differently, which can be seen in the number of data records on SDG 11 concerning sustainable cities and communities.

For comparability between cities, it would be helpful to follow the idea of specialized sample data catalogs, which can be integrated into the existing (or to be developed) open data portals and thus provide a valid, official basis for the calculation of the achievement of objectives. In these catalogs, not only the data itself but also the open data formats to be used as well as rules for the metadata and units for the individual values must be defined. Only then can a calculation and comparisons between cities based on comparable data be made easily. Overall it can be stated that: "[Open Government Data] usage is not simply about uniform barriers that need to be tackled but is a complex process of interaction with technology, communication among stakeholders and government organizations, and collaborative learning" (Ruijer & Grimmelikhuijsen, 2020, p. 14).

References

Attard, J., Orlandi, F., Scerri, S., & Auer, S. (2015). A systematic review of open government data initiatives. *Government Information Quarterly*. https://doi.org/10.1016/j.giq.2015.07.006.

Charalabidis, Y., Loukis, E., & Alexopoulos, C. (2014). Evaluating second generation open government data infrastructures using value models. In *Proceedings of the 47th Hawaii international conference on system sciences, HICSS 2014, Waikoloa, HI, January 6–9, 2014* (pp. 2114–2126). Washington, DC: IEEE Computer Society.

Chen, L. J., Ho, Y. H., Lee, H. C., Wu, H. C., Liu, H. M., Hsieh, H. H., … Lung, S. C. C. (2017). An open framework for participatory PM2.5 monitoring in smart cities. *IEEE Access*, *5*, 14441–14454.

Francisco, A., & Taylor, J. E. (2019). Understanding citizen perspectives on open urban energy data through the development and testing of a community energy feedback system. *Applied Energy*, *256*(April). https://doi.org/10.1016/j.apenergy.2019.113804.

Hartmann, S., Mainka, A., & Stock, W. G. (2019). Innovation contests: How to engage citizens in solving urban problems? In *Civic engagement and politics: Concepts, methodologies, tools, and applications* (pp. 58–77). Hershey, PA: IGI Global.

Klopp, J. M., & Petretta, D. L. (2017). The urban sustainable development goal: Indicators, complexity and the politics of measuring cities. *Cities*, *63*, 92–97. https://doi.org/10.1016/j.cities.2016.12.019.

Lytras, M. D., & Visvizi, A. (2018). Who uses smart city services and what to make of it: Toward interdisciplinary smart cities research. *Sustainability*, *10*(6). https://doi.org/10.3390/su10061998.

Meijer, A. J., Curtin, D., Hillebrandt, M., & Curtin, D. (2012). Open government: Connecting vision and voice. *International Review of Administrative Sciences*, *78*(1), 10–29. https://doi.org/10.1177/0020852311429533.

Meschede, C., & Mainka, A. (2020). Including citizen participation formats for drafting and implementing local sustainable development strategies. *Journal of Urban Science*, *4*(1), 1–19. https://doi.org/10.3390/urbansci4010013.

Meschede, C., & Siebenlist, T. (2019). Open urban data and the sustainable development goals. In *Proceeding of proceedings of the 18th annual international conference on digital government research, June 18–20, 2019*. Dubai, VAE. New York, NY: ACM.

Nugroho, R. P., Zuiderwijk, A., Janssen, M., & de Jong, M. (2015). A comparison of national open data policies: Lessons learned. *Transforming Government: People, Process and Policy*, *9*(3), 286–308.

ODIN. (2020). *Open Data Inventory—Global Index of Open Data*. Retrieved March 27, 2020 from https://odin.opendatawatch.com/.

Open Data Barometer. (2020). *Open Data Barometer*. Retrieved March 27, 2020 from https://opendatabarometer.org/.

Open Data Watch. (2017). *Open Data Inventory 2017 methodology report*. Retrieved March 27, 2020 from http://odin.opendatawatch.com/Downloads/otherFiles/ODIN-2016-Methodology.pdf.

Open Government Partnership. (2020). *Open Government Partnership*. Retrieved March 27, 2020 from https://opengovpartnership.org/.

Open Knowledge Foundation. (2020). *Open Knowledge Foundation*. Retrieved from https://okfn.org/.

Pereira, G. V., Macadar, M. A., Luciano, E. M., & Testa, M. G. (2016). Delivering public value through open government data initiatives in a Smart City context. *Information Systems Frontiers*, *19*, 213–229. https://doi.org/10.1007/s10796-016-9673-7.

Pérez del Hoyo, R., & Mora, H. (2019). Knowledge society technologies for smart cities development. In A. Visvizi, & M. D. Lytras (Eds.), *Smart cities: Issues and challenges* (pp. 185–198). Amsterdam, Netherlands: Elsevier.

Quental, N., Lourenço, J. M., & Da Silva, F. N. (2011). Sustainable development policy: Goals, targets and political cycles. *Sustainable Development*, *19*(1), 15–29. https://doi.org/10.1002/sd.416.

Ruijer, E., & Grimmelikhuijsen, S. (2020). Open data work: Understanding open data usage from a practice lens. *International Review of Administrative Sciences*, *86*(1), 3–19. https://doi.org/10.1177/0020852317753068.

Scarano, V., Cordasco, G., Lettieri, N., Malandrino, D., Manno, I., Palmieri, G., … Rizzolo, D. (2015). Fostering transparency and participation in the data-based society: A Sustainable Architecture for a Social Platform for Open Data. In P. Cunningham, & M. Cunningham (Eds.), *Proceedings of eChallenges e-2015* (pp. 1–9). IEEE.

SDG National Reporting Initiative. (2018). *Facilitating information-sharing on SDG reporting*. Retrieved March 27, 2020 from https://www.sdgreporting.org/.

Sunlight Foundation. (2014). *Open data policy guidelines*. Retrieved from https://sunlightfoundation.com/opendataguidelines/.

Trencher, G., & Karvonen, A. (2019). Stretching "smart": Advancing health and well-being through the smart city agenda. *Local Environment*, *24*(7), 610–627. https://doi.org/10.1080/13549839.2017.1360264.

UN. (2015a). *Goal 16: Promote just, peaceful and inclusive societies*. Retrieved March 27, 2020 from https://www.un.org/sustainabledevelopment/peace-justice/.

UN. (2015b). *Sustainable development goals*. Retrieved March 27, 2020 from http://www.un.org/sustainabledevelopment/sustainable-development-goals/.

UN. (2016). *The new urban agenda*. Retrieved March 27, 2020 from http://habitat3.org/the-new-urban-agenda.

WB. (2018). *The 2018 atlas of sustainable development goals: An all-new visual guide to data and development*. Retrieved March 27, 2020 from http://blogs.worldbank.org/opendata/2018-atlas-sustainable-development-goals-all-new-visual-guide-data-and-development.

WWWF. (2018). *Open data barometer—Leaders edition*. Washington, DC: World Wide Web Foundation.

Zuiderwijk, A., & Janssen, M. (2014). Open data policies, their implementation and impact: A framework for comparison. *Government Information Quarterly*, *31*(1), 17–29.

Chapter 8

The potential of location-based social networks for participatory urban planning

Pablo Martí, Raquel Pérez del Hoyo, Almudena Nolasco-Cirugeda, Leticia Serrano-Estrada, and Clara García-Mayor
University of Alicante, Alicante, Spain

1. Introduction

Human interactions with and within urban spaces have radically changed since social media and mobile devices offer the opportunity to generate content, such as videos, photos, reviews, check-ins, and microblogs, and to share this content with other individuals (Iranmanesh & Alpar Atun, 2018). Indeed, these traces, which reveal how spaces are used and experienced, influence our own way to interact with the environment. Online recommendations, posted images, and, in general, data shared by others about certain activities and urban spaces provide valuable insights about the collective perceived representation and image of places.

This is only one of the main reasons why the volunteered geographic information generated by Location-Based Social Networks (LBSNs) is of special interest for research in the fields of urban studies and planning. It offers up-to-date information about the behavior and human practices that take place in city spaces (Kim, Chae, & Park, 2018). Therefore, LBSNs are possibly the most popular type of online participatory sensing systems (Silva, Horizonte, Salles, & Loureiro, 2013). Data are constantly being generated by millions of people all around the world (Hu et al., 2015), thus representing a potential alternative and/or complementary source to public participation processes for decision-making in urban planning.

Specifically, the crowd-sourced spatiotemporal data obtained from LBSNs has been demonstrated to have a great potential for revealing preferences (Agryzkov, Martí, Tortosa, & Vicent, 2016), opinions (Hermida, 2010), and sentiments (Al Shehhi, Thomas, Welsch, Grey, & Aung, 2019) toward activities, spaces, events, and other related intangible phenomena (Dunkel, 2015) that are

Smart Cities and the UN SDGs. https://doi.org/10.1016/B978-0-323-85151-0.00008-7

not easily detected through information collected by conventional means such as field study observation. Furthermore, these sources allow automatic retrieval, which represents a great advance for urban analysis methodologies, removing constraints related to time and costs associated with traditional methods, for example, interview processes.

There is already a whole body of knowledge that has demonstrated the value of these sources for the design and planning of cities (Afzalan & Muller, 2014; Massa & Campagna, 2014). The application of these data is varied since each social network fulfills certain functions. For instance, these sources allow the continuous monitoring of urban transformations since they can inform about physical and social outcomes. One example of the first case is the physical and permanent outcomes achieved by initiated project actions, in terms of urban economic activity and citizen preferences over existing and newly created public spaces. This could be monitored through social networks such as Google Places, which is the listing of urban activities in Google Maps, and Foursquare, a check-in based social network, i.e., identifying if a food store has opened or closed over the course of the project duration; and if the number of visitors and check-ins has increased or decreased. In the second case, an example could be the covered and uncovered social-related outcomes that may have been achieved over the course of urban transformations, i.e., changes in the type, location, and ranking of Foursquare venues; different spatiotemporal patterns of Twitter activity in urban spaces, etc.

The complementary nature of LBSNs is also an interesting characteristic for urban studies. Their dynamic character can supplement data from static sources, i.e., demographic census; or even complement other types of social media (Gutiérrez Puebla, García-Palomares, Romanillos, & Salas-Olmedo, 2017; Martí, Serrano-Estrada, & Nolasco-Cirugeda, 2019). Silva et al. (2013) have compared Instagram and Foursquare functionalities for studying urban phenomena and social behavior; and Salas-Olmedo, Moya-Gómez, García-Palomares, and Gutiérrez (2018) have adopted data collected from Panoramio, Foursquare, and Twitter for analyzing the digital footprints of urban tourists. These studies have demonstrated three important points: (1) the potential of using different social networks for obtaining a thorough understanding of urban phenomena; (2) a single data source is oftentimes not sufficient to achieve a comprehensive interpretation; and (3) the use of data from various LBSN sources partly offsets the bias that may be present if only one source is used.

In view of the above considerations, this research aims to explore the potential of five location-based social networks—Twitter, Instagram, Foursquare, Strava, and MapMyRun—for obtaining insights about people preferences, opinions, likes, and dislikes about the urban environment that can inform urban planning and management decisions and strategies. Key functionalities and exemplary research using these user-generated information sources are reviewed, highlighting the advantages and disadvantages of adopting these tools for achieving more participatory based decision-making.

Thus, within the context of smart cities (Visvizi & Lytras, 2019), this research is carried out in the framework of the United Nations 2030 Agenda (UN, 2015) to advance the achievement of the Sustainable Development Goals (SDGs), in particular SDG 11 which aims to make urban planning and management more participatory and inclusive. Specifically, the incorporation of geolocated user-generated content from the referenced social networks is increasingly becoming a researching tool, which in turn could help to achieve SDG 11 goals.

2. Location-based social networks for more participatory and inclusive urban planning

In this section, each of the selected location-based social networks is studied in order to identify the opportunities they offer to move toward smarter, more participatory, and inclusive urban planning. After this first approach, some applications will be presented in the following section.

2.1 Social network "Twitter": Spatiotemporal city dynamics

Twitter offers a rich data environment for the analysis of social practices in the city and showcases relevant samples of interactions between individuals and their immediate realm (Frias-Martinez & Frias-Martinez, 2014; Stewart, 2017). This popular microblogging LBSN allows registered users to broadcast text content—with a maximum of 280 characters—and embedded links, photos, or videos with very different purposes. Indeed, Twitter has been proven a useful data source for the development of the most different types of research, from ethnographic to medical studies or human behavior as well as in the urban research field (Martí et al., 2019).

Three meaningful variables contribute to different types of studies in urban planning, such as: (1) the text content, (2) the geolocation of messages, and (3) the timestamp of the tweets. The combination of these variables demonstrates potential on identifying patterns related to people presence, perception, and land use, among others.

Firstly, the representation of geocoded tweets on a map, according to their timestamps, is a straightforward way to observe concentration patterns of people at specific times. This type of analysis showcases people's habits and allows to identify flows of movement happening on a regular basis—during a day or a week—in a certain area or in the whole city (Adnan, Longley, & Khan, 2014; Fujita, 2013). For example, in the case of Alicante city (Spain), 69,763 tweets were found corresponding to a total of 31,666 datapoints after filtering and verifying the data—retrieved from May 23, 2016 to November 8, 2018. The representation of these data on a map shows that specific streets in the city are always concurrent because they concentrate different types of activities developing during working days as well as weekends. However, other areas are frequented more often during working days (Fig. 1. Upper and middle images).

FIG. 1 Tweets in the city of Alicante (Spain). Upper image. Geolocated tweets broadcasted on working days. Middle image. Geolocated tweets broadcasted during weekends. Lower image. Word cloud presenting the most frequently repeated words in the dataset. *(From authors' data mining.)*

This type of representation can be useful when occasional events such as, for example, public demonstrations or cultural festivals happen in the public space. In this latter case, a substantial increase of datapoints in a specific area occurs and fades away once the event is finished (Bolognesi & Galli, 2017; Martín, Julián, & Cos-Gayón, 2019). Secondly, the identification of more frequently repeated words in a dataset and their representation in relation to the place where the tweets were broadcasted allows for the recognition of feelings or perceptions that prevail in each place (Quercia, Aiello, Mclean, & Schifanella, 2015). Moreover, word count techniques can be applied to a Twitter dataset and represented, for example, in a word cloud using scaled text size according to the frequency of words. This is a straightforward technique to overview the most repeated words expressed at a given place that sometimes can coincide with places, festivities, or events in the city, such as happens in the case of the Alicante dataset (Fig. 1. Lower image).

The analysis of Twitter data for urban planning purposes is not without its challenges. Firstly, the analysis of text content can get very intricate depending on the type of research questions that are to be solved (Quercia, Aiello, Mclean, et al., 2015; Saquete, Tomás, Moreda, Martínez-Barco, & Palomar, 2020). Secondly, not the whole of the tweet traffic is geocoded (Murdock, 2011; Sloan & Morgan, 2015) because this requires the activation of geolocation options in the Twitter app as well as the activation of GPS capabilities in the device. In this case, hashtags included in the tweet content are rather useful to infer the approximated geolocation of the messages, or at least a reference of the place where either the tweet has been broadcasted or to which the content refers (Hamstead et al., 2018). And, thirdly, the timestamp that is commonly used to analyze activity patterns—p.e. working days vs weekends—can be modified by occasional events that need to be guessed and considered. In addition, it is to consider that some overactive Twitter accounts may distort the results of any analysis because of the huge concentration of messages in a place or in a specified time.

Despite all these challenges, Twitter data has been demonstrated to be useful for the provision of complementary up-to-date information that shed light on the city dynamics that necessarily are considered in urban planning. Moreover, since it is citizen-generated information, its incorporation into these urban planning processes can be considered an innovative form of citizen participation, in line with SDG 11.

2.2 Social network "Instagram": Studying the city pulse through visual content

Instagram was launched in 2010 as a photo and video-sharing social networking service that also lets users create timeline sequences with their images and videos, as well as those of other users' feeds. Ten years later, this social network has one billion monthly active Instagram users and more than 50 billion photos shared to date (Ali & Aslam, 2020). Currently owned by Facebook Inc., this online social media platform is particularly popular with teenagers and young adults, with 65% of global users aged between 18 and 34 years (Clement, 2020); but evolving from the initial leisure purpose, in the last 5 years Instagram has also become an informal business platform with an increasing 25 million business profiles and more than two million monthly advertisers (Ali & Aslam, 2020). The figures show the embedding of this application in our society and its increasing versatility. From the technological viewpoint, Instagram is continuously updated, introducing new functionalities and innovative new designs, but what makes this social network appropriate for urban researchers? To what extent can it be considered a kind of participatory sensing system to achieve SDG 11?

Among all Instagram features, there are four aspects that make this app suitable for studying urban phenomena and trends based on user's behavior and preferences, namely: (a) popularity, nowadays, it is probably one of the most trendy and hugest banks of geotagged images, which are constantly fed and updated; (b) recognition of user's traces and preferences, due to the nature and volume of Instagrammers' generated information, the identification of hubs of activity in the informal way that they occur in the city's public space is possible, giving insight to urban dynamics on the bases of user behavior more than presumptions; (c) mapping spatiotemporal and social connections, on the one hand, the study of the Instagram knit-network let to identify social clustering, on the other hand Instagram geotagged information permits to represent patterns of encounter in a dynamic time line (Boy & Uitermark, 2016); and (d) multiscale approaches are possible, from spatial large-scale outreach to detailed-scale scopes, as well as variable time-lapse periods according to the research needs. All these issues allow to study the city pulse from a dynamic visual content—user's photographs, providing frequency, time and quantity, and spatial distribution—geotagged location, so unveiling the users' interests and city's rhythm through virtual traces.

2.2.1 Studying the city at local scale with Instagram dataset

An example of the use of Instagram as a tool to assess users' perception after an urban transformation, applied at local scale, is successfully shown in the research carried out by López Baeza, Serrano Estrada, and Nolasco-Cirugeda (2016). The case study objective is the San Francisco Street in Alicante, Spain, a stretch of the city's 16th-century urban tissue that was refurbished in 2013, and popularly nicknamed after the transformation as "Mushrooms Street" in relation to the thematic street furniture installed in this urban setting. This study combines Instagram database information, Google Street view and fieldwork, in order to depict the collective image and the pulse of public life through indicators, such as economic activities before and after an urban refurbishment; the social use of the place's new spatial configuration; and the resulting perceived image of the San Francisco Street (Fig. 2).

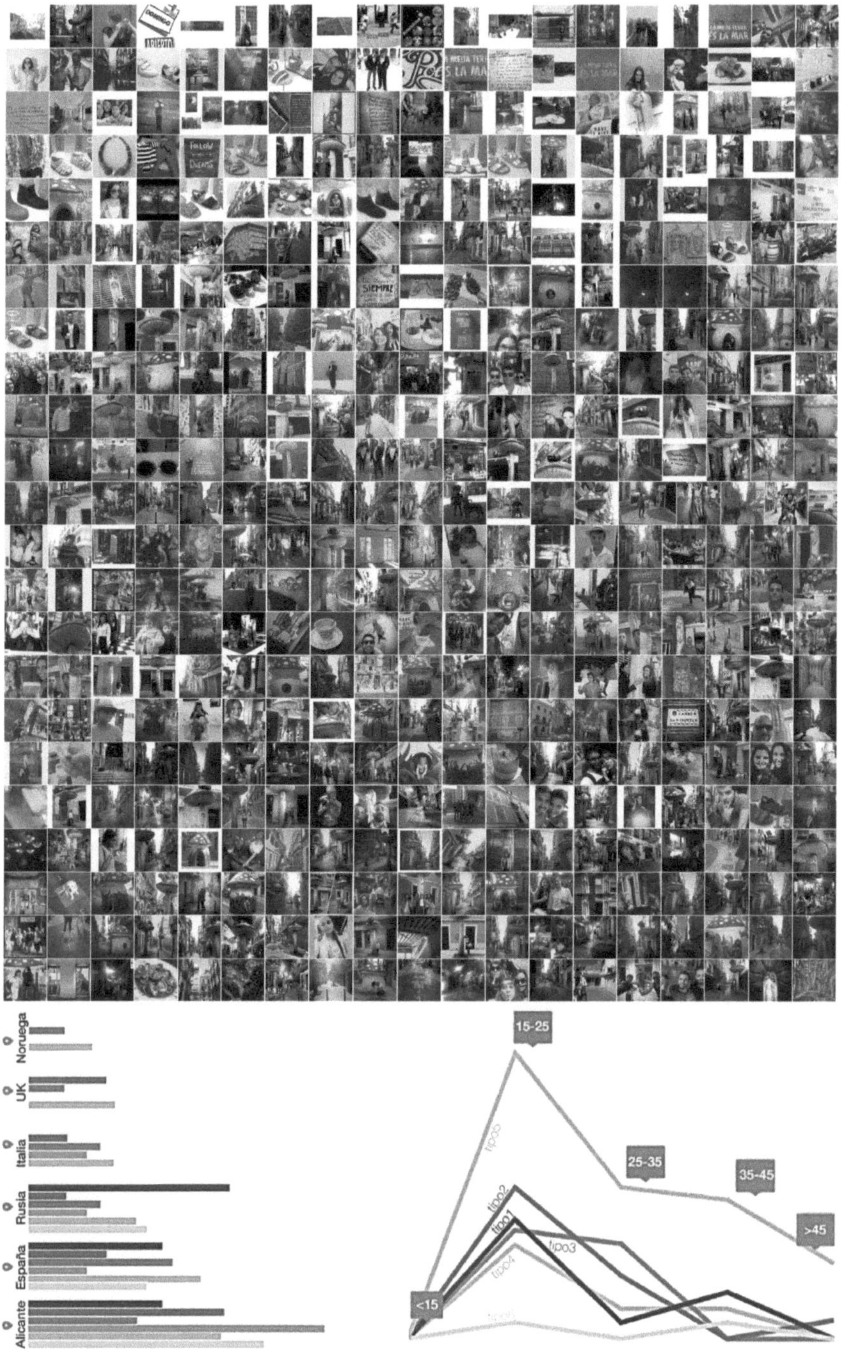

FIG. 2 Users' classification by age range (down right) and nationality (down left) linked to picture's framing and composition (upper part mosaic and lower part graphs' colors). *(Reproduced with permission from López Baeza, J., Serrano Estrada, L., & Nolasco-Cirugeda, A. (2016). Percepción y uso social de una transformación urbana a través del social media. Las setas gigantes de la calle San Francisco. I2 Innovación e Investigación En Arquitectura y Territorio, 4(2). https://doi:10.14198/i2.2016.5.03.)*

Some of the most interesting findings that were obtained with this research are related to the monitoring of people's perception shift. The refurbishment project had taken place in 2013, and the earliest opinions and hashtags after the public space renewal showed a negative perception of the design solution, mainly described as trivial or as a sort of theme park. A year later, in 2014, the general opinion detected was mostly neutral; and finally in 2015, a shift to a positive perception was illustrated in both images and hashtags. Additionally, the same local-scale comments and hashtags were found when researching into the city-scale scope, where this street has become one of the landmarks in the urban scene for locals and visitors (López Baeza et al., 2016).

This study using Instagram geotagged data showed a nontraditional way of detecting urban perceptions and shifts in people's preferences, providing evidence of its potential use in urban studies as a remarkably rich source of information.

2.3 Social network "Foursquare"

Foursquare, created in 2009, is a location-based social network that offers its users the possibility to broadcast their presence in registered establishments and/or points of interest referred to as *venues* by the social network. It also allows registered users to provide insights and opinions about the listed *venues*. For instance, users can rate and recommend *venues*, share pictures, and post "tips" which are comments that often refer to the type of place, activities, and/or details about the experience of the space itself. This is the reason why Foursquare is considered an interesting participatory sensing system (Silva et al., 2013), whose data and metadata are useful for behavioral and place-based analysis purposes.

Foursquare is broadly used worldwide, which has led to its widespread use as a key tool for research in urban studies. By September 2019, it already had more than 55 million monthly active users and 9 million daily *venue* check-ins registered through the mobile phone Foursquare Swarm app (Metev, 2019). The data from Foursquare are accessible, with a limited number of queries per second (Ballatore & De Sabbata, 2020) through its API (Application Programming Interface). Aside from providing a list of the geotagged *venues*, rich and varied metadata associated with each of these can be obtained. Specifically, two types of metadata are worth highlighting since they are often used in urban planning-related research: quantitative and qualitative information.

Firstly, quantitative metadata includes the cumulative number of check-ins and users per registered *venue*. The difference between these two values lies in the fact that the former represents the number of Foursquare users that have checked in at a *venue*, and the latter is the total number of users that have entered a *venue*, regardless of whether they have checked in or not. This quantitative information together is, to a good extent, indicative of the varying degrees of interest and preferences users have for a *venue*. For instance, *venues* within

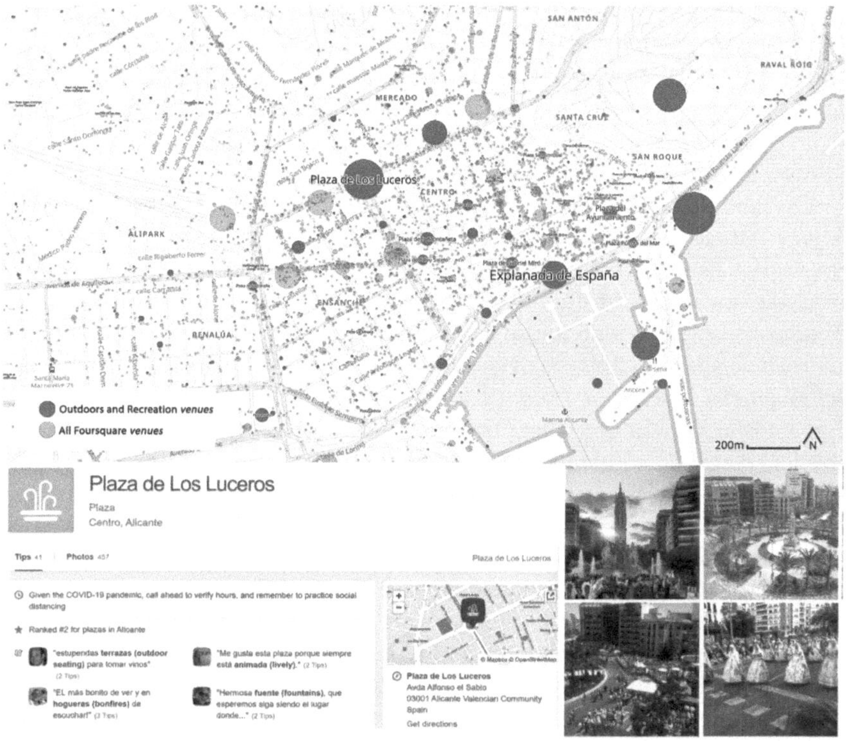

FIG. 3 Upper image, Foursquare *venues* of central Alicante, Spain; lower left and right image, information as displayed in the Foursquare "City Guide" web platform about the *venue* Plaza Luceros. *(From authors' data mining.)*

a given urban area can be ranked by the number of visitors or the number of check-ins, providing clues about the most/less frequented and socially relevant venues (Martí, Serrano-Estrada, & Nolasco-Cirugeda, 2017). An example of this is shown in the Fig. 3 upper image, where all Foursquare *venues* of Alicante city, Spain, are represented with red dots and sized by the cumulative number of check-ins each have received until the date of retrieval, June 1, 2020. Other types of quantitative data that can be used from Foursquare datasets are the total amount of tips and the rating values per *venue* (see, in Fig. 3, lower left image, the 41 Tips and rating for Plaza Luceros).

Secondly, the qualitative information retrieved from Foursquare is twofold. On the one hand, a three level hierarchical categorization of *venues* provides useful and granular information about the type of space, establishment, or activity that takes place in each *venue*. Indeed, when a Foursquare user registers a *venue* in the platform, an important part of the setup is to assign it a specific category from the available predefined categories, subcategories, and/or subsubcategories. The 10 level-1 predefined categories are: Arts and Entertainment;

College and University; Event; Food; Nightlife; Spot; Outdoors and Recreation; Professional and Other Places; Residence; Shop and Service; and Travel and Transport. There are hundreds of level-2 and level-3 categories. For urban related research purposes, the first-level classification is often used to analyze the diversity of *venues* within a given area (Agryzkov et al., 2016; Quercia, Aiello, Schifanella, & Davies, 2015).

On the other hand, the user-generated tips and pictures shared about the *venues* are also interesting qualitative data. These metadata often include Foursquare users' personal recommendations and images and thus perceptual and experiential information about the space. For instance, it is usual to find pictures of *venues* within the subcategory *Plaza*, main category *Outdoors and Recreation*, showing the type of social activities that take place in the space, i.e., the Hogueras festivity in Plaza de los Luceros, Alicante city, Spain, shown in Fig. 3, lower right image. These tips and pictures reveal (i) which activities and physical elements within the space are of collective interests; (ii) the look and feel of the space itself; and (iii) the type of activities and social behavior that can be expected in the *venue*. Given these three aspects, it is possible to affirm that Plaza Luceros in Alicante is a livable and social public space.

When studied together, the qualitative and quantitative data from Foursquare offer great potential for the timely analysis of complex tangible and intangible aspects. The location, distribution, and types of urban spaces and economic activities relevant to the overall urban structure; the social life of the city; and the identity of the urban environment, are among the most common research topics, useful for informing urban planning strategies in accordance with the achievement of SDG 11.

2.4 Sport social networks: "Strava" and "MapMyRun"

MapMyRun and Strava are location-based sports monitoring social networks. Both social networks consist of a website and applications for the main mobile platforms. They allow users to select different sports activities such as running, cycling, or walking. They are mainly based on the location of the sports activities performed by the users, using Google Maps tools and the location systems of the mobile devices.

The main functionalities that these social networks offer to improve urban planning are based on their ability to record and share the geographic data of the places where users pass while performing their activity by means of GPS tracking (Merchant, 2017). In the case of Strava, users can also add multimedia components such as photographs related to the activities and also share these activities through other social networks such as Facebook or Twitter. Thus, Strava's social focus is clearly perceived. This network not only allows the creation of routes that are registered on the Internet, but also promotes community awareness with a clear sense of social interaction. In the field of urban planning, quantitative and qualitative studies are equally important, so it is not possible to take into account only isolated data. This qualitative information enables a better understanding of citizens' needs, preferences, and behavioral patterns (Cook, Shaw, & Simpson, 2016).

From the website of both social networks, it is possible to access and download the routes provided by any user. However, not all these downloads can be done for free. MyMapRun allows users to find existing routes in a given location's area. In addition to the geolocated routes that can be downloaded for free in the GPX format, this social network reports the length of each route and the date it was created, enabling chronological studies. Thus, MapMyRun provides a very useful source of data to identify and analyze the places preferred by citizens to practice their daily sport in the city. Therefore, it is of great interest in the field of urban planning, especially in decision-making processes, enabling the agents responsible for intervening in the city to take into account, as a priority, the needs and preferences of citizens (Laguia Martinez, 2018). Thus, MyMapRun allows the analysis of individual routes created by users, although without providing classifications, rankings, or comparisons when the same routes are also covered by other users.

However, the Strava social network does provide these relationships. Once users finish their sports activity, it is compared and classified with the rest of similar activities performed by other users, identifying the most popular "segments" or stretches of route where users practice sports in the city. These "segments" can be obtained from a previously selected geographical area, and downloaded and analyzed, for example, by means of geographical information systems (Fig. 4).

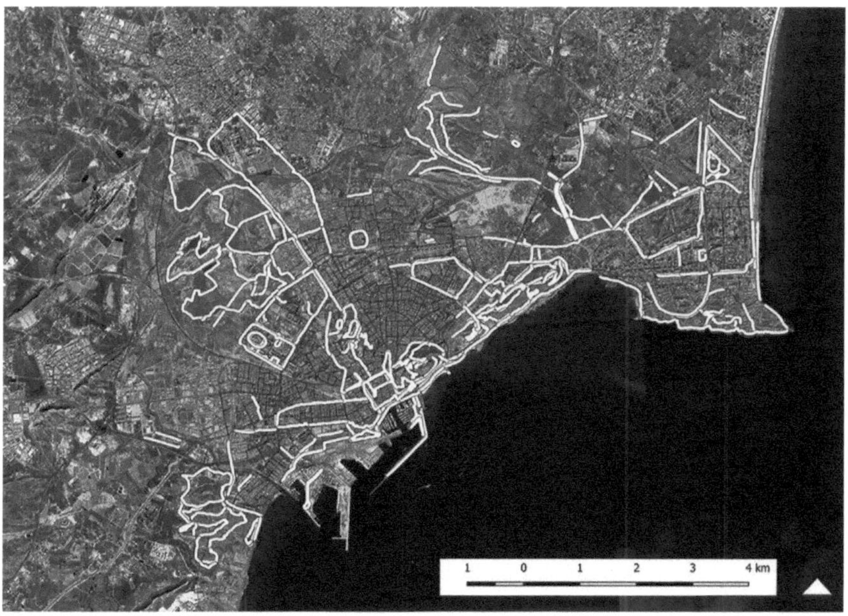

FIG. 4 "Segments" obtained from Strava for the geographical area of the City of Alicante. *(Data from Strava mapped by authors.)*

It is also worth noting the availability of Strava Metro, which analyzes the data that Strava obtains through its application to provide relevant information. However, Strava Metro is not a free service.

In the field of urban planning, it is very important to have data that reflect the greater or lesser degree of sporting activity that takes place in a specific area of the city. This knowledge of citizens' preferences makes it easier for the agents in charge of intervening in the city's decision-making processes (Hochmair, Bardin, & Ahmouda, 2019; Lin & Fan, 2020). In this sense, Strava is a valuable source of data as it allows the downloading of "segments" in which the high intensity of an activity has been previously identified. Thus, Strava offers free downloads of geolocated "segments" in the GPX format and provides information of two types: on the one hand, on the number of users who have performed that "segment" or stretch of route and, on the other hand, on the number of times that all users have performed that same stretch of route. In addition, Strava provides other data on the difficulty of the "segment," its length, etc. The download of "segments," although not complete routes, is allowed free of charge by selecting each of these "segments" from a previously defined graphic window or specific geographic area.

In short, the MyMapRun social network enables urban planners to find out the places and routes that urban sportspeople choose and prefer for their daily sporting activity. From this information, it is possible to characterize these places and establish criteria and priorities for intervention in the city, based directly on the citizens' own experience, i.e., their needs and preferences. On the other hand, the Strava social network provides additional information related to the intensity of use of public space. With these data, urban planners can identify which of all the places preferred by citizens to practice sport in the city are really the most popular. Incorporating this user-generated information into urban decision-making processes can be an innovative new way of understanding public participation and contributing to the achievement of SDG 11.

3. Some examples of good practice applications

LBSN data have been found in many good practice applications for inclusive and participative urban planning as promoted by SGD 11. For example, an analysis of city dynamics was performed by using Twitter to infer spatiotemporal demographics in a city (García-Palomares, Salas-Olmedo, Moya-Gómez, Condeço-Melhorado, & Gutiérrez, 2018). The results presented a deep understanding of the influence of land use on activity patterns during the day. Essentially, geocoded and geotagged tweets were aggregated considering spatial and temporal variables to obtain the number of active users in a defined city zone and time slot. Correlation between active users during the established time slots revealed specific city zones that gained or lost activity over the course of the day. This study provides evidence on city dynamics that were already perceived from observation but was difficult to quantify without the use of these data. These results are relevant for urban planning practice because with these type of studies future patterns of activity in new urban developments can be foreseen according to different land uses; as well as risk assessment or population evacuation, plans can be designed in existing areas of the city according to the expected visitor flow. Moreover, this is of interest for the private sector business activities as the potential demand according to different zones, land uses, and times of the day can be to an extent predicted.

An exemplary application of Foursquare data for achieving inclusive and participative processes has been proposed by the research of Quercia, Aiello, Mclean, et al. (2015) and Quercia, Aiello, Schifanella, et al. (2015). This study has assessed the theory of walkability, which includes the identification of streets that provide safe, interesting, and accessible opportunities for carrying out daily errands. Basically, the study consists of a walkability model based specifically on user-generated data retrieved from Foursquare and Flickr. These two sources have been demonstrated to be useful for pinpointing London's walkable streets, avoiding the need for extensive, manual, and expensive field data collection. Precisely through the analysis of first-level Foursquare metadata categories, it was possible to predict street safety scores by recognizing the presence of specific types of *venues*, thus encouraging specific kinds of street activity. Findings demonstrated that London safe streets are, to a good degree, associated with legitimate activity, such as public basketball courts and community centers and, in general, spatially linked to outdoor open spaces—mainly parks. This exemplary study delivers new insights that are highly relevant for planning and decision-making in line with the SDG objectives for safe cities, by promoting "crime prevention through environmental design" (Newman, 1996).

In relation to the use of Instagram data as a media of better understanding the city's pulse, among the many existing researching works, there are some very interesting studies carried out by Justus Uitermark and John D. Boy from the University of Amsterdam. They found how networks and representations of Amsterdam on social media are "highly uneven" (Uitermark & Boy, 2015). Also, how geolocated images and hashtags from Instagram provide a segmentation of users, highlighting the already existing socio-spatial inequalities (Boy & Uitermark, 2017); or, not to mention their findings about fragmentation and integration among Amsterdam's Instagram users' data, e.g., clusters of "lifestyle enclaves" simultaneously to exchange appraisals by "distant others" that are interpreted as a way of integration between groups (Boy & Uitermark, 2019). These examples show a researching line where the Instagram social network has facilitated a new approach to socio-spatial city analysis as a complementary information for urban planning processes.

On the other hand, a very interesting example of the application of sports social networks is the study carried out by Cook et al. (2016) on the practice of running in the English city of Plymouth. User-generated routes were obtained from MapMyRun. The network also served to record details of dates, times, durations, speeds, description of the running experience, etc. As a result, very interesting conclusions were drawn. For example, how runners develop strategies that enable them to negotiate space with nonrunners; or how runners end up preferring to avoid pedestrian areas precisely so as not to encounter other groups of citizens. Another interesting example is the study conducted by Hochmair et al. (2019) to understand the behavior of the bicycle user community in the Miami-Dade County area. This was done using Strava's tracking data. As a result, the elements of the transportation network, environmental characteristics, and sociodemographic factors that conditioned citizens' decisions about bicycle use were identified. The study used a set of mathematical models to analyze the behavior of a number of complex variables. The conclusions of the study suggest that issues of design of roads and public spaces play a fundamental role in providing greater safety for cyclists. These are just some examples of how geolocalized data from social networks enable the design of urban planning and management processes taking into account the opinions and preferences of citizens, thus being more inclusive and in line with the SDGs.

4. Conclusions

This study suggests that social media data offer unprecedented opportunities for accessing real-time and fine-grained data that can be useful for informing urban planning decision-making processes. Additionally, in some cases, such as Foursquare, data obtained provide a classification into categories and subcategories through presorting and user-specific data processing, which is a helping feature to derive more comparable and precise conclusions.

These new data sources allow to measure, analyze, model, and even predict, in some cases, city dynamics and other phenomena that have always been perceived by observation but measured with difficulty or rigorously represented. Moreover, these new sources contribute to update some methods applied in traditional studies such as the use of public space or labor mobility flows that are commonly analyzed through mobility surveys.

However, it is necessary to consider some drawbacks of using social media data in urban planning processes, such as the requirement of an initial gathering of substantial data; some of the LBSNs presented in this work, e.g., Twitter or Instagram, require thorough and intricate processes of filtering, verification, and classification to become useful for urban analysis; it is a complementary approach to other traditional participatory tools; and studies that adopt these data sources may be aware of and consider privacy and ethical issues.

Although this work has been developed with thematic social networks' sections to better understand the specific features in each case, nowadays, there is an increasing number of studies that combine different location-based social networks to obtain a more complex analysis of a defined area of study, intertwining and complementing several sources.

References

Adnan, M., Longley, P. A., & Khan, S. M. (2014). Social dynamics of Twitter usage in London, Paris, and New York City citation format. *First Monday*, *19*(5). https://doi.org/10.5210/fm.v19i5.4820.

Afzalan, N., & Muller, B. (2014). The role of social media in green infrastructure planning: A case study of neighborhood participation in park siting. *Journal of Urban Technology*, *21*(3), 67–83. https://doi.org/10.1080/10630732.2014.940701.

Agryzkov, T., Martí, P., Tortosa, L., & Vicent, J. F. (2016). Measuring urban activities using foursquare data and network analysis: A case study of Murcia (Spain). *International Journal of Geographical Information Science*, *31*(1), 100–121. https://doi.org/10.1080/13658816.2016.1188931.

Al Shehhi, A., Thomas, J., Welsch, R., Grey, I., & Aung, Z. (2019). Arabia Felix 2.0: A cross-linguistic Twitter analysis of happiness patterns in the United Arab Emirates. *Journal of Big Data*, *6*(1). https://doi.org/10.1186/s40537-019-0195-2.

Ali, H. M., & Aslam, S. (2020). *Instagram by the numbers (2020): Stats, demographics & fun facts*. Retrieved June 6, 2020, from https:www.omnicoreagency.com/instagram-statistics/.

Ballatore, A., & De Sabbata, S. (2020). Los Angeles as a digital place: The geographies of user-generated content. *Transactions in GIS*, *24*(4), 880–902. https://doi.org/10.1111/tgis.12600.

Bolognesi, C., & Galli, A. (2017). Mapping socials a voluntary map of a great event in Monza Park. *Proceedings*, *1*(9), 917. https://doi.org/10.3390/proceedings1090917.

Boy, J. D., & Uitermark, J. (2016). How to study the city on Instagram. *PLoS One*, *11*(6). https://doi.org/10.1371/journal.pone.0158161.

Boy, J. D., & Uitermark, J. (2017). Reassembling the city through Instagram. *Transactions of the Institute of British Geographers*, *42*(4), 612–624. https://doi.org/10.1111/tran.12185.

Boy, J. D., & Uitermark, J. (2019). *Lifestyle enclaves in the Instagram city?*. CUS Working Paper Series—WPS-No. 39 Centre for Urban Studies, University of Amsterdam.

Clement, J. (2020). *Instagram: Age distribution of global audiences 2020*. Retrieved June 18, 2020, from https://www.statista.com/statistics/325587/instagram-global-age-group/.

Cook, S., Shaw, J., & Simpson, P. (2016). Jography: Exploring meanings, experiences and spatialities of recreational road-running. *Mobilities*, *11*(5), 744–769. https://doi.org/10.1080/17450101.2015.1034455.

Dunkel, A. (2015). Visualizing the perceived environment using crowdsourced photo geodata. *Landscape and Urban Planning*, *142*, 173–186. https://doi.org/10.1016/j.landurbplan.2015.02.022.

Frias-Martinez, V., & Frias-Martinez, E. (2014). Spectral clustering for sensing urban land use using Twitter activity. *Engineering Applications of Artificial Intelligence*, *35*, 237–245. https://doi.org/10.1016/j.engappai.2014.06.019.

Fujita, H. (2013). Geo-tagged Twitter collection and visualization system. *Cartography and Geographic Information Science*, *40*(3), 18. https://doi.org/10.1080/15230406.2013.800272.

García-Palomares, J. C., Salas-Olmedo, M. H., Moya-Gómez, B., Condeço-Melhorado, A., & Gutiérrez, J. (2018). City dynamics through Twitter: Relationships between land use and spatio-temporal demographics. *Cities*, *72*(B), 310–319. https://doi.org/10.1016/j.cities.2017.09.007.

Gutiérrez Puebla, J., García-Palomares, J. C., Romanillos, G., & Salas-Olmedo, M. H. (2017). The eruption of Airbnb in tourist cities: Comparing spatial patterns of hotels and peer-to-peer accommodation in Barcelona. *Tourism Management*, *62*, 278–291. https://doi.org/10.1016/j.tourman.2017.05.003.

Hamstead, Z. A., Fisher, D., Ilieva, R. T., Wood, S. A., McPhearson, T., & Kremer, P. (2018). Geolocated social media as a rapid indicator of park visitation and equitable park access. *Computers, Environment and Urban Systems*, *72*(January), 38–50. https://doi.org/10.1016/j.compenvurbsys.2018.01.007.

Hermida, A. (2010). Twittering the news. The emergence of ambient journalism. *Journalism Practice*, *4*(3), 297–308. https://doi.org/10.1080/17512781003640703.

Hochmair, H. H., Bardin, E., & Ahmouda, A. (2019). Estimating bicycle trip volume for Miami-Dade county from Strava tracking data. *Journal of Transport Geography*, *75*, 58–69. https://doi.org/10.1016/j.jtrangeo.2019.01.013.

Hu, Y., Gao, S., Janowicz, K., Yu, B., Li, W., & Prasad, S. (2015). Extracting and understanding urban areas of interest using geotagged photos. *Computers, Environment and Urban Systems*, *54*, 240–254. https://doi.org/10.1016/j.compenvurbsys.2015.09.001.

Iranmanesh, A., & Alpar Atun, R. (2018). Exploring the spatial distribution of geo-tagged Twitter feeds via street-centrality measures. *Urban Design International*, *23*, 293–306. https://doi.org/10.1057/s41289-018-0073-0.

Kim, H. J., Chae, B. K., & Park, S. B. (2018). Exploring public space through social media: An exploratory case study on the High Line New York City. *Urban Design International*, *23*(2), 69–85. https://doi.org/10.1057/s41289-017-0050-z.

Laguia Martinez, D. (2018). Grassroots efforts in contemporary urban mapping: An analysis of alternative uses of collaborative platforms. *Studies of Architecture, Urbanism and Environmental Sciences Journal*, *1*(1), 11–20. https://doi.org/10.22034/saues.2018.01.02.

Lin, Z., & Fan, W. D. (2020). Modeling bicycle volume using crowdsourced data from Strava smartphone application. *International Journal of Transportation Science and Technology*, *9*(4), 334–343. https://doi.org/10.1016/j.ijtst.2020.03.003.

López Baeza, J., Serrano Estrada, L., & Nolasco-Cirugeda, A. (2016). Percepción y uso social de una transformación urbana a través del social media. Las setas gigantes de la calle San Francisco. *I2 Innovación e Investigación En Arquitectura y Territorio*, *4*(2). https://doi.org/10.14198/i2.2016.5.03.

Martí, P., Serrano-Estrada, L., & Nolasco-Cirugeda, A. (2017). Using locative social media and urban cartographies to identify and locate successful urban plazas. *Cities*, *64*, 66–78. https://doi.org/10.1016/j.cities.2017.02.007.

Martí, P., Serrano-Estrada, L., & Nolasco-Cirugeda, A. (2019). Social media data: Challenges, opportunities and limitations in urban studies. *Computers, Environment and Urban Systems*, *74*(September), 161–174. https://doi.org/10.1016/j.compenvurbsys.2018.11.001.

Martín, A., Julián, A. B. A., & Cos-Gayón, F. (2019). Analysis of Twitter messages using big data tools to evaluate and locate the activity in the city of Valencia (Spain). *Cities*, *86*, 37–50. https://doi.org/10.1016/j.cities.2018.12.014.

Massa, P., & Campagna, M. (2014). Social media geographic information: Recent findings and opportunities for smart spatial planning. In *TeMA journal of land use, mobility and environment, special issue: Eighth international conference INPUT Smart City-planning for energy, transportation and sustainability of the urban system* (pp. 645–658). https://doi.org/10.6092/1970-9870/2500.

Merchant, S. (2017). The promise of creative/participatory mapping practices for sport and leisure research. *Leisure Studies*, *36*(2), 182–191. https://doi.org/10.1080/02614367.2016.1231830.

Metev, D. (2019). 41+ must know foursquare statistics in 2020. *Review*, *42*. https://review42.com/foursquare-statistics/.

Murdock, V. (2011). Your mileage may vary: on the limits of social media. *SIGSPATIAL Special*, *3*(2), 62–66. https://doi.org/10.1145/2047296.2047309.

Newman, O. (1996). *Creating defensible space*. Washington, D.C.: U.S. Dept. of Housing and Urban Development, Office of Policy Development and Research.

Quercia, D., Aiello, L. M., Mclean, K., & Schifanella, R. (2015). Smelly maps: The digital life of urban smellscapes. In *Ninth international AAAI conference on web and social media*. Oxford: AAAI Publications.

Quercia, D., Aiello, L. M., Schifanella, R., & Davies, A. (2015). The digital life of walkable streets. In *Proceedings of the 24th international conference on world wide web—WWW '15* (pp. 875–884). https://doi.org/10.1145/2736277.2741631.

Salas-Olmedo, M. H., Moya-Gómez, B., García-Palomares, J. C., & Gutiérrez, J. (2018). Tourists' digital footprint in cities: Comparing big data sources. *Tourism Management, 66*, 13–25. https://doi.org/10.1016/j.tourman.2017.11.001.

Saquete, E., Tomás, D., Moreda, P., Martínez-Barco, P., & Palomar, M. (2020). Fighting post-truth using natural language processing: A review and open challenges. *Expert Systems with Applications, 141*. https://doi.org/10.1016/j.eswa.2019.112943, 112943.

Silva, T. H., Horizonte, B., Salles, J., & Loureiro, A. A. F. (2013). Urban social behavior. A comparison of Foursquare and Instagram to the study of city dynamics and urban social behavior. In *UrbComp '13: Proceedings of the 2nd ACM SIGKDD International Workshop on Urban Computing*. https://doi.org/10.1145/2505821.2505836.

Sloan, L., & Morgan, J. (2015). Who tweets with their location? Understanding the relationship between demographic characteristics and the use of geoservices and geotagging on Twitter. *PLoS One, 10*(11), 1–15. https://doi.org/10.1371/journal.pone.0142209.

Stewart, B. (2017). Twitter as method: Using Twitter as a tool to conduct research. In L. Sloan, & A. Quan-Haase (Eds.), *The SAGE handbook of social media research methods* (pp. 251–265). London: Sage Publications.

Uitermark, J., & Boy, J. D. (2015). Capture and share the city: Mapping Instagram's uneven geography in Amsterdam. In *The ideal city: Between myth and reality. Representations, policies, contradcictions and challenges for tomorrow's urban life* (pp. 1–20). Urbino: RC21 Conference Paper. Urbino, Italy. Retrieved June 18, 2020, from http://www.rc21.org/en/conferences/urbino2015/; (2015).

UN. (2015). United Nations. In *Transforming our world: The 2030 agenda for sustainable development. A/RES/70/1*. New York: United Nations. Division for Sustainable Development Goals.

Visvizi, A., & Lytras, M. (Eds.). (2019). *Smart cities: Issues and challenges: Mapping political, social and economic risks and threats*. Amsterdam: Elsevier.

Chapter 9

The Green Turn: Smart cities, the SDGs, and sustainability in the EU parties' discourse

Cláudia Toriz Ramos
University Fernando Pessoa & CEPESE, Porto, Portugal

1. Introduction

This chapter undertakes the debate on smart cities and sustainability, in the framework of the European Union (EU) party politics discourse and in the context of the European Parliament (EP) elections. Tracing the discussion on smart cities at the level of the EU political debate and governance is rather challenging, because questions on cities imply the consideration of the European multilevel framework of governance, in which the cities fall under the national and subnational layers, the EU having no direct power over them as such. This is to say that the debate has to be indirectly identified, with reference to policies that will have an impact at the level of the city and within the major ideological debate on sustainability and the sustainable development goals (SDGs).

It is therefore an issue area where the virtues of multilevel governance are put to test, since it is of the articulation between the EU and the local and regional levels, upwards and downwards, and with the involvement of the many stakeholders in the process that success or otherwise failure may emerge. Thus, the necessity of involving the citizens takes the ideological and political debate beyond strict utilitarian considerations into the core of democratic governance. The European Parliament, the only directly elected institution of the Union, occupies for that reason a special place in the promotion of democratic debates in the EU context. Political programs and electoral campaigns are, by excellence, the main markers of such discussions. A paradox may nevertheless result from the fact that the local and the supranational layers are the two edges of a complex polity and the cities seem to remain with the former, thus leaving no reason for the EP to be involved. Yet, it may also be that the Parliament, in echoing transnational policy concerns under the many policy areas that impact upon the city, provides a structure of support for fostering and complementing local approaches to the enhancement of the cities. The smart city may that

Smart Cities and the UN SDGs. https://doi.org/10.1016/B978-0-323-85151-0.00009-9

way have something to gain from the transnationalization of politics in the EU. Ideologically, the umbrella debate is presently that on sustainability and its political consequences.

The chapter is divided into two main parts: in the first part the debates on sustainability and the smart cities are briefly introduced, in order to frame the presentation of the European parties' discourse on those topics, in the second part.

2. Framing the debate: Sustainability and the smart city

The centrality of the concept of sustainability in the United Nations' (UN) 2015–30 SDGs agenda brought the environmental concerns forward and mainstreamed them across all policy areas. Furthermore, because environmental issues are of a clear transnational dimension and only scarcely limited by political borders the agenda has to be tackled as one, for the entire world, specificities then having to be decided with reference to the circumstances of each territory.

The origins of the debate on sustainability in the UN date back to the Brundtland report (Brundtland Commission, 1987), where the concept of sustainable development was defined as "development that meets the needs of the present without compromising the ability of future generations to meet their own needs" (Part I, Chapter 2, paragraph 1). The concept of "needs" was presented not as a rigid minimum but as a sociocultural construct. However, the report stressed the idea that sustainability is not just about intergenerational equity but also about intragenerational equity, thus inviting the world to address poverty and lack of resources as major reasons for imbalance and underlying condition for ecological and other crises (Part I, Chapter 2, paragraph 4). This way, social issues were also connected with sustainability, which means it also went deeply into the internal balance of each and every human society. The report specifically addressed urbanization (Part II, Chapter 9) where concerns with growing third world cities, with industrial world cities, with housing for the poor, with local governance and citizens' involvement, and with international cooperation, among others, were already present. Governance was also acknowledged as a necessary tool for promoting sustainable development. The text also mentioned the regional (supranational) dimension and its capacity to coordinate in between countries in order to foster common agendas (Part III, Chapter 12, paragraph 31).

The problem is therefore not new. However, if there is a major change, from the time the report was issued to the present, it certainly is the speed at which technological innovation evolved, so as to make possible solutions that were not foreseeable as such at that time. The smart cities movement is one such case, or another way put, it can be normatively framed within the debate on sustainability and development, not least because it has managed to bridge technological tools with concerns of ecological and equitable development, and with social justice and democracy (Visvizi & Lytras, 2019). In 2015, under goal 11 of the SDGs, "Make cities and human settlements inclusive, safe, resilient and sustainable," the UN took the opportunity to carefully detail targets for the urban world, among which some touch the smart cities debate namely, goal 11.2 on sustainable transport systems, goal 11.3 on inclusive and sustainable urbanization, which includes participatory processes, or goal 11.6 on air quality and waste management (UN, 2017).

As expressed in the title, this chapter takes as political focus a regional entity, the European Union, while aggregating layer of a multilevel polity, which also encompasses the national and the subnational levels of its 27 member states, and which is an interesting case for the regional (supranational) approach. The EU has indeed acclaimed the SDGs agenda and seeks to bridge its policies with the guidelines therein set (European Commission, n.d.-a). With reference to the UN 2015 agenda, however, the Union was already pioneering in some respects (e.g., environmental policy) and thus, at least some of the UN guidelines come as reinforcement of previously set regional guidance. For the specific purposes of goal 11, the European Commission (n.d.-b) monitors a series of initiatives that meet the targets and provide indicators for that goal. The internal action of the EU has to be addressed mostly under the common regional and cohesion policies and the common environmental policies, with which energy and transports policies bear a major connection (cf. Latoszek, Visvizi, Paszkowski, & Wojtowicz, 2021). Externally, the EU extends its promotion of goal 11 by resorting to the enlargement and neighborhood policies and to international cooperation. While the latter are remarkable fields of action and of research, the scope of this chapter is only the internal (intra-European) dimension (European Commission, n.d.-b).

The "Urban Agenda for the EU" was created with the Pact of Amsterdam, in 2016 (Council of the European Union, 2016). It is not a policy, it is a new and multilevel "working method" for the cooperation between member states, cities, the European Commission, and other stakeholders. It aims to stimulate growth, liveability, and innovation in the cities and to tackle social change (European Commission, 2019a, n.d.-c). For the purpose, several key areas of action were identified (from air quality, to digital transition and urban poverty, to name but a few) and a specific focus on better regulation, better funding, and better knowledge was set. An urban development network of cities and urban areas was created in order to reinforce exchanges and for implementing integrated actions based on sustainable urban development strategies and financed by ERDF, the European Regional Development Fund (2014–20) (European Commission, n.d.-d).

In the framework of the environmental policies, the enhancement of the sustainability of the Union's cities was defined as a priority in the *7th Environment Action Programme* (2014–20) (Decision No. 1386/2013/EU of the European Parliament and of the Council of 20 November 2013) as follows:

In order to enhance the sustainability of Union cities, the 7th EAP shall ensure that by 2020:

(a) a majority of cities in the Union are implementing policies for sustainable urban planning and design, including innovative approaches for urban public transport and mobility, sustainable buildings, energy efficiency, and urban biodiversity conservation.

This requires, in particular:

(i) agreeing on a set of criteria to assess the environmental performance of cities, taking into account economic, social, and territorial impacts;
(ii) ensuring that cities have information about, and better access to, financing for measures to improve urban sustainability;
(iii) sharing best practice between cities at Union and international level in relation to innovative and sustainable urban development; and
(iv) in the context of ongoing Union initiatives and networks, developing and promoting a common understanding of how to contribute to improved urban environments by focusing on the integration of urban planning with objectives related to resource efficiency, an innovative safe and sustainable low-carbon economy, sustainable urban land use, sustainable urban mobility, urban biodiversity management and conservation, ecosystem resilience, water management, human health, public participation in decision-making, and environmental education and awareness.

More recently, the new European Commission for the 2019–24 mandate, under the leadership of Ursula von der Leyen, has clearly elected the "European green deal" (climate neutrality for Europe by 2050) and the digital transformation as political priorities (European Commission, 2019b). The Commission plays a major role in developing EU political strategies, in designing and implementing EU policies and even in fostering transnational networks across the Union for bridging its many layers for common causes. However, it is for the European Parliament candidates to face elections, in order to gain direct legitimacy. It is therefore in their hands, rather than in the Commission's, the tricky task of communicating with the citizens. The following section thus covers the ideas relating to sustainability, and implicitly the smart cities' debate, as expressed by the main political groups of the EP and associated European political parties, especially at the time of the 2019 election.

3. European parties' discourse: Smart cities and sustainability

The momentum for broad ideological and political debate in the EU happens every 5 years, at the time of EP's elections. However, it is widely known that the campaigns are too often a lost opportunity for a truly European-oriented debate, frequently falling into national issues, if not petty politics (Kreppel, 2002; Raunio, 2017). It is not different with issues concerning the cities and urban life, or even rural communities (Visvizi, Lytras, & Mudri, 2019), which may all be perceived by the citizens as a topic of the concern of local authorities and local elections only, rather than a set of issues with European-level implications. The bridges between urban policies and regional development, cohesion policies, energy and transports, agricultural and environmental policies, technology, and innovation can and should be established throughout the campaign, so that the citizens perceive the supranational layer as also a relevant one in the many aspects of their daily local life, and so that the EP could also benefit from local democracy and participation.

Ideologically, the political debate on the environment is well established though, and has even taken increased visibility in the public sphere in the recent years. At the time of the 2019 elections it was under the spotlight, because of the "kids for climate" movement. Climate change is a regular topic in European debates and in the media, the 2015 Paris Agreement and more generally "sustainability" as buzzword having filled substantial space in the discussions (European Commission, 2020).

For the public, "ecology," "environment," and "green politics" are concepts often used interchangeably, because they depart from a common concern with the environment. However, in deeper ideological debates, they do not exactly target the same thing. Whereas ecology is more often connected with a "radical" approach to environmental degradation and to the necessity of reintroducing balance in the humans-nature relation, by substantially changing the system of production and the ways of life, environmentalism, also reckoning the problems, will nevertheless put forward political solutions based on reforms and not on the complete subversion of the status quo. It can be argued that ecologism tends to ecocentrism, while environmentalism remains anthropocentric and technocentric. In this context, it is obviously rather difficult to imagine the role of cities, even if smart, in a world led by a substantial "return to nature" (humans as a species among many) as sometimes claimed by radical ecologism but, to a certain extent, the debates on the smart city and on the smart villages are coterminous (Visvizi et al., 2019; Visvizi & Lytras, 2019). Green politics entered the political scenario back in the 1970s for designating those political parties and programs that reckoned the necessity to act upon environmental imbalance. They were initially strongly anti-system parties, but in the meantime some managed to reach governmental offices, what led to some mitigation of their radicalism but also to increased incorporation of environmental concerns in mainstream policies (Barry, 2014; Dobson, 2007).

Public perception of those nuances may not be so established though. Nevertheless, a change has happened in recent years. While in the past, there were sectors of the population and political parties that could be characterized as "environmental deniers," more recently the environmental tide has been such that it is difficult for a political party to remain on stage opposing, ignoring, or simply avoiding the topic. There was therefore a "green turn" in politics (e.g., European Commission, 2019c, 2020). The research thus assesses the impact of the "green turn" in European party politics. The following subsections present and compare the main tenets put forward by the European political parties and families, within the broad ideological and policy areas that bridge urban and sustainability issues. Whether environmental and related urban questions are addressed at all and whether they are addressed as a peripheral or a structural issue, are two major questions guiding the inquiry.

TABLE 1 European political parties and related European Parliament Groups.

European political parties	Parliamentary groups in the European Parliament	
	Name and 2019 electoral results (%/seats)	
European People's Party (PPE/EPP)	*Group of the European People's Party* (Christian Democrats) (PPE)	24.2/182
Party of European Socialists (PES)	*Group of the Progressive Alliance of Socialists and Democrats in the European Parliament* (S&D)	20.5/154
Alliance for Liberals and Democrats for Europe Party (ALDE)	*Renew Europe Group* [former Group of the Alliance for Liberals and Democrats for Europe (ALDE)]	14.4/108
European Democratic Party (PDE/EDP)		
European Green Party	*Group of the Greens/European Free Alliance* (Greens/EFA)	9.9/74
European Free Alliance (EFA)		
Identity and Democracy Party	*Group Identity and Democracy* [former Group *Europe of Nations and Freedom* (ENL)]	9.7/73
European Conservatives and Reformists Party (ECR)	*European Conservatives and Reformists Group* (ECR)	8.3/62
European Christian Political Movement (ECPM)		
Party of the European Left	*Confederal Group of the European United Left—Nordic Green Left* (GUE/NGL)	5.5/41

There are at present 10 political parties registered with the Authority for European Political Parties and European Political Foundations (n.d.). They are trans-European parties, the members of which are mainly national parties. They are connected with the EP parliamentary groups, but not necessarily on a one-two-one basis, some groups gathering support from more than one political party and some parties splitting their elected members by more than one group. The list of parties considered and their main affiliation with the EP groups is presented in Table 1. The sequence of their presentation follows the results of the 2019 election, in order to provide evidence on their effective representation in the European Parliament, i.e., number of MEPs and, therefore, number of votes. The left-right axis may also be useful for interpreting the data, given the structural connection between anticapitalism and radical ecologism. However, a caveat is to be made on the usual broad convergence at the center in EU politics, as opposed to the edges, both right and left, where antisystem discourses normally find shelter (Kreppel, 2002; Raunio, 2017).

4. European People's Party

In its programmatic presentation, the European People's Party (EPP) connects the safety of the citizens with "high environmental and social standards." The EPP also states: "We believe in the potential of green and sustainable growth in a social market economy based on the principle of environmental sustainability" (EPP, n.d.). Their view is therefore one that brings together a market economy with concerns of sustainability and high social standards.

For the purpose, they think the European Union needs to tackle water supply, environmental degradation, and loss of biodiversity, by contributing to the "conservation and regeneration of natural resources (air, water, and soil)." Besides, they also focus on "renewable energy sources and phasing out unsustainable energy production." They thus acknowledge climate change as a challenge. They affirm the capacity of the EU to be the front-runner in the cleantech sector and the necessity to progressively reform the transport sector, presented as fundamental for global trade. The party expresses support for the EU Emission Trading System, for the reduction of emissions, and, in order to reduce dependency on fossil fuels, for the development of "a more integrated, cost-effective EU mechanism to support renewable energy technologies which are not yet competitive." The concern is with the "protection of industrial sectors facing fierce international competition" (EPP, n.d.).

The 2019 "elections Manifesto" (EPP, 2019) again acknowledges the relevance of environment—"a clean planet to our children and grandchildren"; but is also emphatic on digital innovation and world competition—"We must reconcile protecting the environment and biodiversity with a dynamic economy." For the purpose, it supports the Paris agreement, and emphasizes that the EU can actually pioneer in the new economy of sustainability and in "e-mobility." The aim is to reinforce investment and innovation in sectors leading to "technological solutions to low-carbon mobility." It also stands for a "true Energy Union, where electricity and gas can cross borders without any obstacle, to bring down citizens' living costs, support the cost-effective expansion of renewable energies, and make sure no external power can undermine Europe's energy security." Therefore, it addresses energy from the point of view of both security and liveability. The regional and local dimensions emerge in tandem with this, since it is stated that the party wants to overcome gaps between regions, and between cities and rural areas, by fostering "connection and mobility across all regions of Europe," by means of a strong cohesion policy: "We will boost investments into rail, road, and digital infrastructure to make every region competitive." It is further stated that they want to combat intra-EU brain drain "by guaranteeing access to high-speed internet and world-leading supercomputers in all European regions." The same cohesion purpose is expressed in aiming to provide "telemedicine to rural communities" and in planning investment in "technologies to support the development of 'smart' cities and villages to increase the quality of life in all parts of the continent." The emphasis on digital tools is very strong: citizens' digital literacy and digital competences, the potential of artificial intelligence and robotics, biomedicine, big data, and mobility engineering are all mentioned. This focus is tied together with being competitive in the world markets, in particular with reference to China and the United States. The Manifesto mentions the necessity to ascribe common funds to these aims, in order to create a true "European Innovation Union" (EPP, 2019). Overall the focus is strongly economic.

5. Party of European Socialists

In the Declaration of Principles of the Party of European Socialists, it is stated: "Our renewed vision of solidarity is a joint investment in our common future. It means lasting justice and solidarity between generations. It means we preserve the planet, protect the elderly, and invest in young people" (PES, 2011). The wording very much echoes the Bruntland report conception of sustainability, by bringing together social justice claims (intra and intergenerational solidarity) and environmental concerns. The same social and environmental focus is kept through the text, notably by declaring that "values-driven growth means that environmental sustainability, human dignity, and well being are fundamental to wealth creation" and that it entails a "new economy," where the public sector has an important role to play. The same happens in the 2019 electoral Manifesto, where they propose for the EU a "Sustainable Development Pact with social and ecological targets to ensure that economic interests do not trump the environment" (PES, 2019).

European leadership in world sustainability is also stated, together with the necessity for a "Just Transition Fund" to help implement the UN's SDG Agenda with social fairness. The Paris Agreement is supported. Hence, European industrial renewal and the use of climate neutral renewable energies, by "making polluters pay and supporting investment in affordable clean energy" are deemed necessary. Improving mobility, air quality, and affordable energy-efficient housing are also set as targets for fostering citizens' quality of life. In this context, the setting of a "Plan for Affordable Housing and Clean Public Transport in Europe" is proposed. In line with this focus on liveability, the Manifesto also states that the PES will "fight energy and water poverty, and guarantee accessible, healthy and good quality food, safe products and an end to damaging exposure to toxic chemicals." Although the PES and the EPP share the broad European political center and often find their member parties in office, sometimes in coalition, the rhetorical emphasis of the socialists is comparatively stronger on the social dimension of the policies, whereas EPP clearly shows a major concern with market efficiency (PES, 2019).

6. Alliance of Liberals and Democrats for Europe Party

The third political group in the EP gathers the liberals of the Alliance of Liberals and Democrats for Europe Party (ALDE) and the European Democratic Party (EDP). Liberal standpoints have been the matrix of this group which, with the previous two, has been from the outset represented in the EP.

The ALDE party stands for "sustainable development and peace in the world," but the text of the program does not draw a clear connection between both. It is nevertheless stated that a "safer and just society" will enhance peace in the world (ALDE, n.d.). In the 2019 EP electoral manifesto, the party identifies, among the challenges of the present: "climate change, environmental degradation, plastic pollution, and an urgent need for a clean energy transition" within a list that goes well beyond strictly environmental concerns. Throughout the document, the concern with the "transition to a resource efficient circular economy" is stated along with a strong focus on meeting the terms of the Paris Agreement and the achievement of the "energetic transition." The latter plays a key role in the arguments presented by the party, which also points out the necessity for a "reduction of industry and transport emissions including aviation and shipping." At this point, the text evolves into standing for innovation notably "greater investment in smart mobility, cross-border, and sustainable infrastructure projects including Trans-European Networks, in particular for railroads." The support for the EU emissions trading system is also stated. Among the challenges, the Manifesto also identifies: "the revolutionary impact of technology, especially artificial intelligence and questionable handling of personal data," but there is no specific mention to the smart cities debate. From the point of view of governance, the text emphasizes the importance of international cooperation for environment, thus multilateralism. For the implementation of the new transport networks, it mentions the importance of the EU cohesion policy for fostering "sustainable economic growth" (ALDE, 2018).

7. European Democratic Party

The European Democratic Party (EDP) asserts that "Climate change is no longer a theory, it is a reality. It is no longer about 'future generations' but it affects our current environment." The party also mentions the "cruel paradox" that the effects of global warming affect more heavily the least polluting countries (EDP, n.d.). Reckoning all this, they propose a series of measures to transition to a "green economy," which they present as the "third industrial revolution." Among those are: an ambitious renewable energy policy, a strategy for European self-sufficiency and energy security, putting a price on carbon, and establishing a tax on financial transactions for the creation of a global investment fund for the reduction of greenhouse gas emissions. They believe a common energy policy is necessary in the EU, focusing on "energy renovation of buildings, smart grids, energy storage systems, and renewable energy production." They also propose measures to address problems concerning well-being and liveability, notably "vegetalization programs in urban areas," or the creation of a "European civil protection force." The Manifesto also emphasizes the necessity to protect biodiversity, which connects with agriculture policies (EDP, 2019).

8. European Green Party

The European Green Party was born in 2004, as a follow-up of the European Federation of Green Parties. Quite logically environment issues are central in their foundational charter, where they state:

> *Greens recognize that the values of environmental responsibility, freedom, justice, diversity and non-violence can be – and are to some extent – shared by other political families. What defines Greens is that we consider these values as interdependent and inseparable.*
>
> (European Greens, 2006)

This notion of politics as an ecosystem that has to balance human and environmental concerns is reinforced by their stating that: "beyond any notion of utility, Greens believe that each of the diverse species of life on our planet has an intrinsic value and beauty and therefore deserves to be protected." In addressing the concept of "Environmental Responsibility," they state: "Taking responsibility for our biosphere is a central tenet of Green values. Society depends on the ecological resources and the health and resilience of the planet, and we bear an over-riding obligation to future generations to protect this inheritance." Further down in the text they affirm: "We advocate strongly the need to live within our ecological means." This is then connected with the protection of biodiversity and the sustainable use of renewable resources (European Greens, 2006).

Ecological imbalance is acknowledged: "Our European pattern of production, consumption, and commerce are contributing to the continued poverty of the majority of the planet's inhabitants and causing severe environmental degradation and climatic instability." Its structural dimension is pointed out: "The urgent need to change these patterns necessarily means a profound adjustment process if we are to reverse this damaging exploitation of our common home," further stating: "The political challenge facing us is the task to restructure the global agenda so that economic and trade policies serve social and environmental objectives and not just economic indices." For the purpose, global cooperation and intense European cooperation are deemed necessary (European Greens, 2006).

The Greens interlink climate change and energy issues. They stand for the Paris Agreement and for "a future without fossil fuels and nuclear energy." For them, "the priority is to develop decentralized and renewable alternative energies." In this context, they emphasize again that,

> *For us, climate action is not a nice addition to other policy content to appear ecological whenever it is deemed useful. Instead, it is the very substance of our political struggle and we will keep on being uncompromising on the necessity of concrete climate policies.*
>
> (European Greens, 2020a)

They connect this debate with transport by stating the necessity to "increase energy efficiency and promote sustainable ways of transportation," notably public transport and shared mobility (European Greens, 2020a). The same concern on sustainability is expressed on environment and food: "Living with, not against nature. We stand for the respectful treatment of the environment and animals, and support sustainable means of food production." A mention to "consumers' rights to sustainable and healthy food" is added to this: "Good food is everyone's right. We stand for shorter food chains that link consumers and producers together in fair regional partnerships." They frame this debate within the concern with "caring about the wellbeing of mankind, animals, and nature," a formula that resonates with ecological approaches (European Greens, 2020b).

In the 2019 electoral Manifesto, the Greens state that they want "Europe to lead the world in a just transition to sustainable societies." The program thus encompasses climate action measures, notably energy transition, leaving behind fossil fuels and nuclear power. On transports, they propose reducing the necessity for transport, and transitioning to cleaner means of transport—railways instead of airplanes. On urban issues, they propose the creation of "low-emission zones in towns and cities across Europe" and the promotion of "cycling and walking." Environment and nature should be protected by moving into sustainable farming and fishing, and by significantly expanding protected areas. It is also stated that "animals have the right to live free from abuse" and that "no public money should go to industrial animal farming." The pattern of a "circular economy" is advocated. A new organization is to be set: "Workers and regions need a just transition to sustainable livelihoods. A special European scheme should be set up to finance retraining and moving to new jobs, providing social security and alleviating fears" (European Greens, 2019). The approach of the Greens to the environment is therefore more radical, even close to ecologism, since they do not see a way forward without major structural changes to present days' life styles.

9. European Free Alliance

The European Free Alliance gathers parties generally representing subnational regions of some EU member states. Their points of view are for that reason quite interesting for the debate on cities and on urban-rural relations. In the EP, they sit mostly with the Greens.

Part 4 of the European Free Alliance Manifesto for the 2019 elections, is called "Building a Sustainable World" (EFA, 2020). There, the party states:

"The principles of growth and development that underpin our modern society have led to the depletion of resources, climate change and a loss of biodiversity across the planet." They therefore think that it is necessary to limit the impact of human activities on the environment and to protect European citizens from all environmental risks to their health. Their regional imprint is quite visible when they state: "More efforts should be made to reconcile the interests of important national and regional agriculture and fisheries industries with the EU's climate change goals." They further state that "Innovation in regional sectors can boost sustainable and organic agriculture, fisheries and rural development and lead to a more comprehensive regional policy." They also stand for energy efficiency and renewable sources, and support an EU "ambitious energy policy." They are against fracking, for the risk of ground-water contamination and against nuclear power. From the point of view of energy governance, they favor "more energy sovereignty" for the regions and the nations and broad participation of the subnational levels in the decision-making processes. The connection of sustainability with decentralization thus emerges from their discourse (EFA, 2020).

10. Identity and Democracy Party

To the right of the political arch of the EP stands the group Identity and Democracy backed by the party carrying the same name, but also known as Movement for a Europe of Nations and Freedom. The party was founded in 2014, and the parliamentary group was renamed after the 2019 election. They are a fairly large far right group, but also a fairly recent one. Therefore, the website of the party does not contain information on the topics addressed for the purposes of this chapter and that is the reason why the group's ideas on the topics here debated are not included. Clearly they are not among the priorities of the group, but it may well be that this does not mean unawareness of environmental issues (Identity and Democracy Party, 2019).

11. European Conservatives and Reformists Party

Not as far right as the previous, but standing on the conservative and eurosceptical corner of the EP the European Conservatives and Reformists Party (ECR) and the European Christian Political Movement (ECPM) together back the European Conservatives and Reformists Group in the EP. The ECR party shows awareness of the environmental issues, but coupled with market and capitalism. They state: "how we can best improve our environment using market mechanisms; the innovation and entrepreneurship that comes from capitalism, and the natural sense of stewardship and trusteeship that is part of Burkean conservatism." In line with this, they emphasize the role of "corporate and individual responsibility in environmental issues." With reference to the ideological alignments, they state: "Environmentalism, and the world we live in, are altogether too important to be left to the left." In this context, the party also shows concerns with deforestation, rewilding, and clean energy mix. In sum, they state that "Conservatives are natural conservationists" thus drawing connection between political conservatism and the conservation of nature (ECR, 2019).

12. European Christian Political Movement

The European Christian Political Movement (ECPM) "has a relational view of economy and understands the purpose of all economic activity as supporting life and advancing wellbeing for all." In their 2019 electoral Manifesto, they state that industry and agriculture should be addressed, in the EU context, in close connection with sustainability. They advocate for "green sustainable industries" and want to ensure that "the polluter pays." They also acknowledge the importance of the protection of environment as an EU policy topic and under the intergenerational solidarity clause. They reinforce the importance of technological innovation to protect the environment and renew energy production. With reference to transportation, they invoke the principle of subsidiarity to legitimize decentralized options on choices to be made (ECPM, n.d.).

13. Party of the European Left

On the other edge of the political spectrum, the Party of European Left is less emphatic on environmental issues than in general the other parties previously mentioned. For instance, they focus on energy, but substantially from the point of view of "energy poverty," a situation still characterizing a significant part of the European population, according to them. This is therefore a social not an environmental approach to energy issues (European Left, 2018). However, under its working groups' structure, the party does tackle the issue of environmental change, but once again the focus is very much a social one, what the party calls a "left-wing concept of a Green New Deal." This is presented as a social-ecological transformation and is based on a radical critique to capitalism, subversive, but anthropocentric (European Left, 2020):

> *At a time when climate change is posing an existential threat to the planet we need to radically transform our economies and societies. It characterizes the left that it connects ecology and social issues. We therefore speak of the need for a social-ecological transformation. Necessary is a 'system change' that goes beyond capitalist modernization. With the climate movement we share a common criticism of the prevailing climate damaging doctrine of the capitalist system and the position that industrial production based on fossil energies has no future.*

14. Conclusion

What do the programs and electoral manifestos of the EU parties tell us about sustainability and the smart cities? They confirm the "green turn," i.e., that core items of the debate can no longer be ignored and that the consensus in reckoning the existence of the sustainability problem is quite broad. This is valid for climate change, for the necessity to step into renewable energies, for technological innovation, for fighting poverty, and for sustainable urban as well as rural environments.

The same consensus is not so obvious when it comes to actual solutions. Whereas some parties state the necessity of implementing reforms, while keeping a sharp eye on markets and global competition, others advocate for a clearly social imprint tying together environmental sustainability and social responsibility. Still others will say that the problem is structural and, therefore, that there is a major necessity for system reconversion. The succession of standpoints presented goes from right to left. To the far left, social-led anticapitalist reconstruction is advocated, whereas to the right the conservation of nature may go in tandem with conservatism in general. Subnational approaches reinforce the role of decentralization and potentially of more participatory processes. The green party is, among the European parties, the one that plunges more deeply into the ecological arguments, bringing together human beings with other living species under an ecological ecosystem to be reconstructed, even if at the cost of major changes.

Where do the cities, the smart cities stand, in the middle of the debate? Due to decentralization tenets and the subsidiarity principle, the cities fall outside the direct scope of the supranational polity and, therefore, of the explicit political EU parties debate. However, they are profoundly inside, once the economy, transports, energy, open spaces, food, liveability, regional development, technological innovation, or, as a whole, ecological balance start being debated. Whether this is a hindrance or rather an opportunity for the cities remains to be proven, possibly on an empirical basis. Ideologically and legally the EP sets the framework for the cities of the future, without explicitly politicizing the local debate, which is beyond its political limits. However, the citizens voting in local elections and for regional and national parties are the very same citizens voting for the EP. The coherence of the whole, without being a rule, may well become a tendency. Furthermore, democratic, notably participatory practices, at the local level are much easier to foster than at the supranational level and are actually definitional for the smart city (Ramos, 2019); but democracy in a multilevel system must be the all-encompassing regime or it will not work. Hence, raising the citizens' awareness of the profound intersectionality of environmental, urban, regional, energy, transports, and rural policies, will pave the way for a more enlightened debate and for the citizens recognition of the practical (and potentially beneficial) effects of a multilayered structure of governance.

Through the looking glass of the EP parties' manifestos, the city of the future emerges: cleaner, closer to the nature, relying on efficient and clean transports and energy, more equitable, humanized and technological, self-relying and democratic, traditional or renovated, the choices depend on the voting options of the citizens. The offer is there and the political debate public.

References

ALDE—Alliance of Liberals and Democrats for Europe Party. (2018). *Freedom, opportunity, prosperity: The Liberal vision for the future of Europe*. Retrieved from https://www.aldeparty.eu/sites/alde/files/40-Resolutions/2019_freedom_opportunity_prosperity_the_liberal_vision_for_the_future_of_europe_0.pdf.

ALDE—Alliance of Liberals and Democrats for Europe Party (n.d.). Political programme and European elections manifestos. Retrieved from https://www.aldeparty.eu/political-programme-and-european-elections-manifestos.

Authority for European Political Parties and European Political Foundations (n.d.). Homepage. Retrieved from http://www.appf.europa.eu/appf/en/authority/welcome.html#.

Barry, J. (2014). Green political theory. In V. Geoghegan, & R. Wilford (Eds.), *Political ideologies: An introduction* (pp. 153–178). London: Routlege.

Brundtland Commission. (1987). *Our common future*. Oxford: Oxford University Press.

Council of the European Union. (2016). *Urban agenda for the EU. Pact of Amsterdam*. Retrieved from https://ec.europa.eu/futurium/en/system/files/ged/pact-of-amsterdam_en.pdf.

Dobson, A. (2007). *Green political thought* (4th ed.). London: Routledge.

ECPM—European Christian Political Movement (n.d.). An economy that works for people and planet. Retrieved from https://ecpm.info/campaign/an-economy-that-works-for-people-and-planet.html.

ECR—European Conservatives and Reformists Party. (2019). *Blue-Green Summit V*. Retrieved from https://www.ecrparty.eu/event/ecr_party_blue_green_summit_v.

EDP—European Democratic Party. (2019). *Election manifesto—Europe stand up!*. Retrieved from https://2019.democrats.eu/en.

EDP—European Democratic Party (n.d.). Energy and climate. Retrieved from https://www.democrats.eu/en/theme/energy-and-climate.

EFA—European Free Alliance. (2020). *Manifesto 2019: 4. Building a sustainable world*. Retrieved from https://www.e-f-a.org/2019-manifesto-european-free-alliance/4-building-a-sustainable-world/.

EPP—European People's Party. (2019). *Manifesto 2019*. Retrieved from https://www.epp.eu/files/uploads/2019/05/EPP-MANIFESTO-2019.pdf.

EPP—European People's Party (n.d.). Environment Retrieved from https://www.epp.eu/our-commitments/commitments/.

European Commission. (2019a). *Urban agenda for the EU. Multi-level governance in action*. Brussels: European Commission.

European Commission. (2019b). Communication from the Commission to the European Parliament, the European Council, the Council, the European Economic and Social Committee and the Committee of the Regions. In *The European green deal*. Brussels: COM. 11.12.2019. 640 final.

European Commission. (2019c). *Special Eurobarometer 490 April 2019 "climate change"*. Brussels: European Commission DG COMM.

European Commission. (2020). *Special Eurobarometer 501 March 2020 "Attitudes of the European citizens towards the environment"*. Brussels: European Commission DG COMM.

European Commission (n.d.-a) Sustainable development goals. Retrieved from https://ec.europa.eu/info/strategy/international-strategies/sustainable-development-goals_en.

European Commission (n.d.-b). Goal 11. Make cities and human settlements inclusive, safe, resilient and sustainable. Retrieved from https://ec.europa.eu/sustainable-development/goal11_en.

European Commission (n.d.-c). Urban agenda for the EU. Retrieved from https://ec.europa.eu/futurium/en/urban-agenda.

European Commission (n.d.-d). Urban development. Retrieved from https://ec.europa.eu/regional_policy/en/policy/themes/urban-development/network/.

European Greens. (2006). *Charter of the European Greens*. Retrieved from https://europeangreens.eu/node/5745.

European Greens. (2019). *Manifesto 2019*. Retrieved from https://europeangreens.eu/priorities-2019-what-european-greens-fight#manifesto.

European Greens. (2020a). *Climate & energy*. Retrieved from https://europeangreens.eu/positions/climate-energy.

European Greens. (2020b). *Environment & food*. Retrieved from https://europeangreens.eu/positions/environment-food.

European Left. (2018). *Right to energy*. Retrieved from https://www.european-left.org/campaigns/right-to-energy/.

European Left. (2020). *Environment*. Retrieved from https://www.european-left.org/working-groups/environment/.

Identity and Democracy Party. (2019). *Homepage*. Retrieved from https://www.fr-id-party.eu/.

Kreppel, A. (2002). *The European Parliament and supranational party system: A study in institutional development*. Cambridge: Cambridge University Press.

Latoszek, E., Visvizi, A., Paszkowski, M., & Wojtowicz, A. (2021). Poland and the energy sector under transition: Imperatives, actors, issues. In A. Visvizi, A. Matysek-Jedrych, & K. Mroczek-Dąbrowska (Eds.), *Poland in the single market: Politics, economics, the euro*. London: Routledge. https://doi.org/10.4324/9780429285301.

PES—Party of European Socialists. (2011). *PES declaration of principles*. Retrieved from https://www.pes.eu/en/about-us/our-values/pes-declaration-of-principles/.

PES—Party of European Socialists. (2019). *A new social contract for Europe. PES Manifesto 2019*. Retrieved from https://www.pes.eu/export/sites/default/.galleries/Documents-gallery/PES-Manifesto-2019_EN.pdf_2063069299.pdf.

Ramos, C. T. (2019). Democracy and governance in the Smart City. In A. Visvizi, & M. D. Lytras (Eds.), *Smart cities: Issues and challenges. Mapping political, social and economic risks and threats* (pp. 17–30). Amsterdam: Elsevier.

Raunio, T. (2017). European parties: A powerful caucus in the European Parliament and beyond. In D. Hodson, & J. Peterson (Eds.), *Institutions of the European Union* (4th ed., pp. 357–379). Oxford: Oxford University Press.

UN—United Nations. (2017). *Work of the Statistical Commission pertaining to the 2030 agenda for sustainable development*. A/RES/71/313, 10 July 2017.

Visvizi, A., & Lytras, M. D. (Eds.). (2019). *Smart cities: Issues and challenges. Mapping political, social and economic risks and threats*. Amsterdam: Elsevier.

Visvizi, A., Lytras, M. D., & Mudri, G. (Eds.). (2019). *Smart villages in the EU and beyond*. Bingley, UK: Emerald Publishing.

Part III

Safety, security, exclusion/inclusion, well-being

Chapter 10

Exploring the relationship between the smart-sustainable city, well-being, and urban planning: An analysis of current approaches in Europe

Vasiliki Geropanta, Anna Karagianni, Sofia Mavroudi, and Panagiotis Parthenios
Technical University of Crete, Chania, Greece

1. Introduction

Progressive urbanization, climate change, and the economic crisis are only some of the contemporary global challenges that have forced cities to redefine their public spaces and manage their urban health and citizens' well-being (Capolongo et al., 2018). Current research on spatial design increasingly focuses on how cities have become central to securing a sustainable future (UN, 2015), with the United Nations 2030 Agenda for Sustainable Development pioneering strengthened efforts to ensure the health and promote well-being for all (UN, 2017). Because urban living is associated with risk factors for developing both physical and mental health disorders, the search for innovative solutions, easy to access and human in perspective, has become urgent (Fraga, 2019).

Despite the vast publications and popularity that the topic of well-being has generated, the challenges of urban living are massive for citizens. For example, poor air quality, transportation issues, noise pollution, waste management, inconvenient access to open spaces as well as noncommunicable diseases that include mental health (UN-Habitat & WHO, 2020) are only some of the challenges to address. In this ambiguous environment, many contemporary urban actors, policymakers, planners, and research drivers all work to explore methods that improve living conditions and mitigate the risks from the aforementioned causes of stress (Alohan & Calvo, 2020; Capolongo et al., 2018; Corburn et al., 2020; Pineda & Corburn, 2020; Pineo et al., 2020).

Smart Cities and the UN SDGs. https://doi.org/10.1016/B978-0-323-85151-0.00010-5

In these efforts, sustainable urban development and smart cities are among the development strategies that seem to dominate the discussions for a number of reasons all related to their ability both to enhance city life from a user's point of view and to make it more competitive in a global world. In certain cases, the two work in parallel thus intersecting to create the "smart sustainable city model" (Angelidou et al., 2017). In this model, technology seems to gain the lion's share among the methodologies of mitigating city risks; accurate, consistent, and timely data facilitates the service provision in cities and assists in the efforts for a better quality of life (Angelidou et al., 2017). The popularity of this model can also be understood since a glance at the UN's research indicates this suggested intersection (UN, 2019).

However, although these fields have been continuously enriched since 2015 (Visvizi & Lytras, 2018a, 2018b, 2019a, 2019b) the connection between the smart sustainable city and the achievement of well-being has not yet revealed its full potential. Many current types of research highlight the importance of exploring the relationship between sustainability and the smart city at a deeper level (Bifulco, Tregua, Amitrano, & D'Auria, 2016), while others emphasize the need to approach sustainability and smart cities from a more practical point of view (Angelidou et al., 2017). Still, others search for a focus on design and architectural concepts (Bibri & Krogstie, 2017). But a search on these topics in the mainstream journals reveals that very few studies have been published about the intersection of the three in explaining how the term of smart city contributes to the overall concept of prosperity.

Considering this research gap, this specific work aims to explore how the applications of smart cities can potentially promote urban wellness that targets a sustainable way of living. This objective is addressed through a benchmark analysis of the literature review, and an in-depth investigation of three well-known case studies concerning smart city applications oriented to that goal. Specifically, the case studies aim at illustrating not only the state of the art in achieving well-being but also how this objective responds to the policy "Mind the GAPS framework" (UD/MH, 2016), which connects urban planning strategies with sustainability and the well-being sector. The results can be used as a basis for further exploration of the well-being indicators in the urban context, as well as the development of smart city applications that improve city dwellers' quality of life.

In light of the above, this chapter is structured as analyzed below: firstly, we explore and identify the concept of a smart sustainable city and urban health to explain the relationship between urban environments and mental/physical well-being. In addition, we intend to reveal all the factors that affect this relationship. This specific approach assists in investigating the well-being framework and in exposing the weaknesses of contemporary urban/public space to respond to a more healthy way of living. This part aims also at connecting the discussion with spatial planning and highlights the urban elements that can be considered as modifiers in achieving well-being. Next, we present our case studies, thus revealing and highlighting the potential bridge that the two paradigms contribute to the field. Lastly, the results allow us to identify smart city strategies that enhance or create well-being factors.

As such, this chapter contributes to the popular debate about smart cities and SDGs in a dual way: on the one hand, it investigates the timeless but also very up-to-date subject of promoting city wellness, and on the other hand, it brings to light some of the cities that have already adopted well-being concepts as one of their priorities. Lastly, the chapter contributes to broaden the discussion by enhancing the research around the concept of smart city and actually locating the smart city initiative within the framework of "the 'civil society,' bearing its own power separated from the state" (Bassi, 2017; Bouskela, Casseb, Bassi, Luca, & Facchina, 2016).

2. Literature review

2.1 The smart-sustainable city model

The intersection of the two city models (the smart and the sustainable) has revealed itself only in the last few years. A large amount of research that focuses on sustainability within smart city initiatives has engendered definitions, all as a result of pressure from the sustainability movement in urbanized cities (Ahvenniemi, Huovila, Pinto-Seppä, & Airaksinen, 2017; Bibri, 2019; Cucuzzella & Goubran, 2019; UN, 2019). Among the many important contributions of these definitions is that sustainability is achieved through the use of online tools and ICT. This occurs when cities monitor urban functions using real-time data "to become more flexible, responsive, and resilient" (Angelidou et al., 2017). In this body of literature, there are three ways to approach the convergence of smart and sustainable city models (Angelidou et al., 2017). The first one works in a way that sustainability is the target, while smart is the instrument. The second one works with the two terms as if they are of the same force toward the city smartness, smart city planning, and smart growth (Angelidou et al., 2017). The third one is the framework in which the two models work, with the smart city model working more toward social sustainability as opposed to the environmental one. Lastly, it is worth mentioning here the "zero vision" campaign as opposed to people (zero CO_2 emissions). This campaign describes methodologies to achieve minimum or no "traffic incidents, zero waste, and zero aggression in public spaces" (Komninos, 2016; Zaman, 2014).

In addition, new research about smart city development has begun focusing on the express feelings of its urban residents brought about by the strategic planning for the city (Guedes, Alvarenga, Goulart, Rodriguez, & Soares, 2018; Haarstad & Wathne, 2019; Lin, Zhao, Yu, & Wu, 2019). In these cases, the goal of smart city construction has expanded to include well-being, evaluating it as an important parameter to achieve city smartness (Lin et al., 2019). What is still missing is a clarification of the way health and well-being are integrated into the smart city agenda and how this shifts the focus toward the achievement of citizens' empowerment and inclusion. This factor turns out to be an important objective for smart cities meaning to work both with physical but also mental and social well-being (Trencher & Karvonen, 2019).

2.2 Well-being—Urban health in contemporary urban settings

Urban health is a current subject in urban planning studies covering fields beyond illness and disease to include mental and social well-being (Thomas, Sabel, Morton, Hiscock, & Depledge, 2014). It can be defined as the study of "urban characteristics"—meaning the material and immaterial essence of the surrounding environment and its infrastructure—elements that can influence health and disease in the urban context (Ompad, Galea, & Vlahov, 2017). The theoretical raison d'être stems from the realization that although the quality of life has generally improved worldwide, mental health problems have increased, particularly in developed industrialized countries (WHO, 2019). As highlighted by the "Centre for Urban Design and Mental Health" (UD/MH), higher rates of mental illness occur in cities than in rural areas, as city dwellers show higher rates of depression, and other similar diseases, because of the way they live. This said, it seems that mental health disorders can adversely impact the health and wellness of communities worldwide and vice versa. There is a significant amount of proof that links the urban environment with mental health (Evans, 2003).

In accordance with the above, some research reveals that stress and urban sickness occur when there is "an imbalance between environmental demands and human response capabilities" (Evans & Cohen, 1987, p. 28). Others blame urban living for reasons that are inherent to urban conditions, arguing on their connections with stress-related factors (Litman & Victoria Transport Policy Institute, 2020). For example, environmental stress can generally be regarded as a combination of individual and physical factors. However, the most important items seem to be noise, overcrowding, and fear of crime, social isolation and loneliness, light pollution, toxic pollution, inadequate physical activity, inadequate access to nature, among others (Litman & Victoria Transport Policy Institute, 2020).

From a spatial point of view, recent research reveals the role of architecture and urban planning in achieving urban health (McCay, 2017). For example, some point to how urban green areas impact our mental health and the sense of well-being positively (Coldwell & Evans, 2018; Hunter et al., 2019; Vujčić, Tomicevic-Dubljevic, Zivojinovic, & Toskovic, 2019; Wood, Hooper, Foster, & Bull, 2017). Specifically, nature has a restorative potential in reducing stress when it comes to citizens accessing or just looking at urban parks and natural settings (Hajrasouliha, Rio, Francis, & Edmonson, 2018). Similarly, other research asserts that some urban settings can have a restoration potential that is comparable to or more noteworthy than natural environments which are to be found in the concept of attractivity (Herzog, Maguire, & Nebel, 2003; Hidalgo, Berto, Galindo, & Getrevi, 2006). They found that the most attractive urban environments were generally either "historic-cultural" or "recreational places," in terms of aesthetic and restorative qualities. Lastly, expanding green spaces can be further regarded as a crucial spatial framework for networking and for creating a sense of belonging. In this way, this sector builds on social interactions, social inclusion and therefore reinforcement as the key to achieving social well-being (De Silva, McKenzie, Harpham, & Huttly, 2005; Hajrasouliha et al., 2018).

Indeed, it seems that cities invest in this sector looking at the places and the ways to experiment (Maller, Townsend, Pryor, Brown, & St Leger, 2006). Redesigning or revisiting or simply even discussing public open spaces (POS), for example, has become very crucial, as is the case with many policies, from the environment to wellness and happiness. This poses the advantage that cities engaged in POS encourage human interactions and therefore indicators of well-being are more easily measured by looking at the impact on well-being through these occurrences (UN-Habitat, 2019). Besides other functions, public space also encompasses a symbolic interpretation since it also conveys special cultural meanings (Carr, Francis, Rivlin, & Stone, 1993), which all add to the quality of space.

SDG 11 focuses on the factors that make up the quality of life in the city, arguing that services and public spaces must be affordable, of good quality, and well distributed in the urban fabric to guarantee territorial equality. According to Francis, Wood, Knuiman, and Giles-Corti (2012), "POS quality appears to be more important for mental health than POS quantity" so developing high-quality POS may most probably help communities become more resilient in these diseases.

2.3 The well-being factor in EU policies: Four steps for urban planners

In their early stages, urban design and urban planning seem to start by discussing the well-being sector through a number of policies that aim at bridging the two. "Mind the GAPS Framework" is a tool that identifies four topics as key and tangible axes for the design or evaluation of existing urban spaces and cities in terms of their prospects of integrating the term wellness into their design. It has been abbreviated as a methodological framework and was published by the "Centre for Urban Design and Mental Health" in May 2016 after being reviewed by a wide range of experts. It consists of four pillars—"green places, active places, prosocial places, and safe places"—acting as a set of goals for the urban designers and policymakers concerned with the sustainable development of mental health. This framework collects and categorizes all the factors that contribute to well-being; hence, in the context of this research, it becomes the means of evaluating whether or not smart city strategies adopted by some cities have managed to promote well-being.

G/green places: The search for accessible green areas is vital to people's daily lives as they maintain and improve mental health and well-being.

A/active places: Integration of physical activity in people's daily routines, either by promoting active ways of transport or by endorsing the creation of outdoor sports facilities is of major importance due to the fact that they improve the general state of mood and well-being.

P/prosocial places: The design and development of urban public spaces with active citizen involvement is crucial as public spaces promote interactions among different groups of people and therefore perform as social spaces.

S/safe places: The role of safety and security is separately highlighted inside this framework. An integrated city planning that prevents potential urban hazards (traffic, environmental pollutants, and a sense of getting lost) is pivotal to improving people's mental health and well-being (Fig. 1).

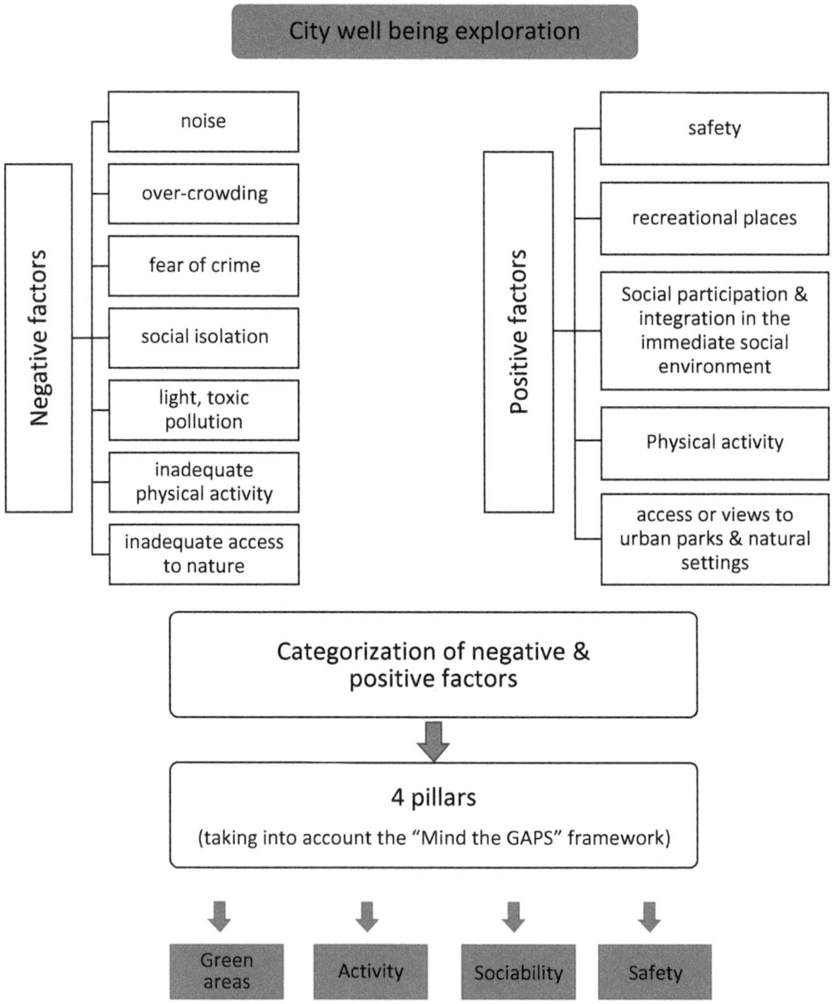

FIG. 1 Exploration of the positive and negative factors that affect city well-being and categorization into four pillars taking into account the proposed "mind the GAPS framework." *(Based on authors' elaboration.)*

3. The research project

The project consists of an investigation of three European cities (Barcelona, Copenhagen, and Vienna) with the objective to reveal how their smart city initiatives promote well-being and its response to the scope of "Mind the GAPS Framework." Based on the chosen UDMH's (Urban Design Mental Health) methodology assessment, four crucial pillars are explored, namely "green places, active places, prosocial places, and safe places." Therefore, the first part of each case study relates to the definition of "smart and sustainable," as extrapolated through the applied city strategy. Then each case study is highlighted regarding the methods that bring about urban well-being by focusing on the four categories.

3.1 "Smart" Barcelona and the role of Superblocks in building the well-being sector

The City of Barcelona is one of the leading smart cities in Europe, using ICTs as a way of improving urban functions for more than a decade (Angelidou, 2016). With dispersed projects running in various departments, the city's current priority is to evolve the concept of smartness by taking advantage of highly transformational data-driven technologies to promote the three key axes of its strategy: transform the city, achieve innovation, and empower people.

As formulated in the "Barcelona Digital City" plan, the city's overall strategy focuses on investing in digital infrastructures (urban technologies) for the general management and the assurance of city residents' public needs. Urban technologies conduce to improve citizens services in order to confront the biggest urban challenges (housing, energy, mobility, etc.), to which health is included through the collaboration among all actors: "government, industry, academia, and citizens" (Angelidou, 2016). These synergies are facilitated through ubiquitous connectivity, open data, and living labs, which encourage citizens' involvement and participation. In this sense, it can be included in the second category of smart-sustainable city, as explained above.

Among all projects, "Superblocks" reconfigure the real/natural space of the city. As one of the most representative cases of urban innovation projects led by the Barcelona City Council program, "Superblocks" are square territorial units of up to nine city blocks that are restrictive in-car use. This urban development creates a dynamic inner area in which pedestrianization of the streets, new space for public transport, and bicycle routes, as well as space for new uses, are encouraged. Thus, through "Superblocks," public space is reclaimed for the public by endorsing sustainable mobility and promoting social cohesion and collaboration. In addition, "Superblocks" aim to increase green areas and areas for outdoor activity aligning with the long-term vision of smart city Barcelona that plans to double tree coverage throughout the city and increase space for parks. New green spaces and the released public space leave room for outdoor markets and other outdoor activities and as a result, new meetings and recreation areas are being formed.

From a technological point of view, the implementation of the "Superblocks" model in Barcelona's strategy explains how projects that build on existing assets of the city through city monitoring can achieve safety, a better quality of life, welfare, efficacy, and competitiveness in various socioeconomic aspects with ICT. "Superblocks" intervention not only establishes more physical activity (walking and cycling) into citizens' daily routines but also strengthens the factor of safety in the whole city by discouraging car use since this action leads to a lower risk of car accidents. To achieve these goals, the star smart city applications "Senti lo," "Bicing," and "Smart Citizen" seem to be the most effective.

Firstly, "Senti lo" is an innovative open-source software platform that through an open network of sensors distributed throughout the city of Barcelona collects and depicts information regarding the energy consumption of public buildings, urban noise, temperature, and pollution levels as well as the flow of people, bicycles, and cars. Real-time measurable indicators of the city's current state are accessible by all, assisting citizens to adapt their activities according to their comfort and safety. Secondly, the "Bicing" service app aims at updating the list of bike stations, to better manage bike availability and to optimize vehicle transit, through constant collection of urban data. As an app, it promotes bicycle use and therefore citizens' physical activity. To this goal, a supportive action of a point system has been integrated in order to reward users with access to products and services related to healthy living, ecology, and sustainability. Thirdly, "Smart Citizen" is an open-source platform for the monitoring of environmental indicators with the use of crowdsourcing techniques, generated by participatory processes. It relies upon the distribution of the "Smart Citizen Kit," which is an urban sensor kit developed by Fab Lab Barcelona and Hangar that measures environmental-related values such as air quality, temperature, sound, humidity, and light quantity. Through the same kits, users can upload information from the immediate urban environment and share the measured values in real time. Therefore, databases with values addressed to the comfort and safety well-being factors are generated and available for all.

3.2 Smart, sustainable, and technological Copenhagen

In Copenhagen, the functional sectors that have been selected to work all together with climate, mobility, livability, and urban development support the smart city's ambitious vision to become the world's first carbon-neutral city by 2025 (De Falco, 2019). Their goals include achieving a cleaner environment, healthier citizens and a higher quality of life (The City of Copenhagen, 2019). Copenhagen is working toward the SDG's goals through coordinated internal efforts but also external city collaborations such as C40 in order to disseminate local solutions worldwide. The relationship between urban development, citizen involvement in urban procedures, and the aspiration for a sustainable city have led to the transformation of the city into a living lab which serves as an ideal test bed for gaining useful knowledge and experience of smart city IT solutions.

In the smart city implementation, "Copenhagen Solutions Lab" works with innovative partnerships to foster intelligent technologies and data-driven solutions (The City of Copenhagen, 2015a).

The "Street Lab" established by Copenhagen Municipality 2016 is an urban laboratory located in the city center, where innovative solutions (such as smart parking solutions, urban greenery, waste management techniques, and measurement of air quality indicators) are tested under real urban conditions. It develops data to support solutions that reduce air pollution in Copenhagen, reduces citizens' exposure to air pollution by creating urban solutions to avoid it, and examines how air pollution is transmitted to buildings. Other than that, it develops data for "urban nature" care, which is a strategy that aims to create more urban nature and enhance the quality of the existing nature in Copenhagen. Therefore, the above actions can be regarded as highly promoting the factors of green spaces and safety in terms of well-being. The general objective of the city is to examine various solutions for their future possibilities and their constraints regarding their implementation to larger areas of the city.

A supportive action to the City of Copenhagen's vision to "Co-create Copenhagen, vision for 2025" is the publishing of annual "Urban Life Accounts." Although it is not explicitly stated in the strategy as an initiative for the smart city, it is developed under the broader concept of promoting livability for the city. This is achieved by gathering various data extracted from questionnaire surveys, field surveys, and other means about the factors that influence urban development such as pedestrian mobility patterns and observations made from the experience of the city on a human scale with the goal of providing evaluations about the level of general urban quality of life in the city (The City of Copenhagen, 2015b). The effort to measure, map, and annually monitor the status of everyday urban life trends is a meaningful initiative in enhancing well-being aspects as it gives feedback so that the development of the city meets the needs of its citizens.

The city of Copenhagen has recently been developing the "PLUSnet system" for cyclists and according to the city's "Bicycle Strategy" for the time period 2011–25 aims to expand the cycle track network and increase the share of the bicycle network that has three lanes in each direction to 80% by 2025 (The City of Copenhagen, 2012). The whole system consists of green routes, bicycle superhighways, and local bicycle routes and integrates the initiative of "green wave" cycle flow improvement. More specifically, through ITS (Intelligent Traffic Systems) traffic lights are coordinated to allow cyclists to maintain an increased, even speed, and reduce the number of stops on their ways. Bicycles (when more than five) are being prioritized when approaching an intersection and thus contribute to a continuous flow of traffic. Traffic adjustment, user interaction, and continuous monitoring contribute to its ongoing development (Centre for Public Impact, 2016a, 2016b).

Regarding well-being, the "PLUSnet system" aims to increase the percentage of the population that chooses bicycles as the main means of daily transport to work and school. Prioritization of bicycles and improved cyclists' travel times promote the factor of physical activity for all users while safer intersections and continuous maintenance contribute significantly to improving other well-being factors such as a better level of comfort and safety. The variety of speeds on chosen cycle routes makes room for diversity so that many cyclists can follow the travel tempo that best suits them, also promoting the level of interaction among users and therefore sociability. As an overall goal, the improved bicycle culture is beneficial for the quality of public space and the city's atmosphere (Centre for Public Impact, 2016a, 2016b).

3.3 Smart-sustainable City Vienna (Austria)

Smart City Wien is a comprehensive strategy for transforming the City of Vienna in the areas of sustainability, digitalization, and participation. Its key objectives are the quality of life, resource conservation and innovation which all together contribute to sustainable urban renewal and therefore to the achievement of the UN climate goals. In the Smart City Wien Framework, the subject areas have been categorized into buildings and digital services. In the above domains, users can select their range of interests, relevant applications, tools, and smart city services. For each subject area in the Smart City Wien Framework Strategy, objectives and indicators are defined, monitored, and evaluated for their level of achieving the goal (Vienna Municipal Administration, 2019a, 2019b).

One characteristic project of "Smart City Wien Framework Strategy" that closely addresses urban development and the real/natural space of the city is the project "Greening the city with plant tubs." It's about the development of the low-tech "BeRTA Green Facade Module" that started as a research project for installation in 50 demo greenhouses as an exclusively street-side greening facade to be placed in public spaces. The pilot implementation of these facades together with an innovative web-based participation and application instrument for detouring the long approval processes (thus making it easily implementable in buildings facades), as well as the continuous monitoring of measurable social and technical effects of the greening of buildings, make it an integrated solution for which a high demand has already been noted. The city of Vienna has already created funds for its further and rapid implementation across the city. Related to the city well-being, planting greenery on large building facades reduces the temperature thus preventing urban heat islands and improving the urban microclimate both in the buildings as well as in the immediate surroundings. As a result, there is more comfort and safety while walking in the city. Besides, it improves air quality and helps the creation of new green views inside the densely populated urban areas, both of major importance and potential regarding the well-being of the citizens and ensuring a healthier environment in the city (Vienna Municipal Administration, 2019a, 2019b).

A lot of interesting smart city applications are developed under the healthcare domain of Vienna's smart city framework ranging from mobile health applications, assistance systems, home care monitoring aimed to either provide data about nursing, or care services or promote preventative health. Regarding the incorporation of digital applications to augment city well-being in public space, two paradigms have been selected.

The "Wien gibt Raum" (Sharing Public Spaces) digital service captures and maps information about Vienna's public space and urban furniture with technologies such as mobile mapping and imaging technologies. With the use of geodata services and GIS systems, it further provides an organizational framework for the better management of objects, activities and traffic flows in public space. The new public space management thus facilitates bureaucratic procedures with effective networking between various city departments. Furthermore, it provides a holistic measure of keeping public spaces usable and accessible for more citizens. This novel and integrated approach also aim to declutter urban public space by making it more accessible and coherent for all user groups.

The "Smart traffic lights" is a networked traffic light system that uses recognition algorithms for persons' movement patterns and thus improves the general experience of walking in the city. In addition, the traffic light system is equipped with weather and environmental data sensors that measure temperature, humidity, and pollutants. Interconnected traffic lights provide constant data that are integrated into GPS navigation systems to optimize traffic flow and reduce emissions. All these apart from creating valuable databases, contribute to strengthening comfort and safety factors of residents' city well-being (Vienna Municipal Administration, 2019a, 2019b) (Table 1).

4. Conclusions

This specific study sheds light on the well-being potential in three European cities because of the implementation of their smart-sustainable city strategies. Firstly, a number of important observations preestablish the criteria under which the smart city strategies are investigated in order to build well-being in each city although vaguely implied within the scope of the studied urban strategies. This point is important since the pursuit of well-being is not an—one at a time—intervention method but rather a process that is slowly revealed in these strategic efforts. Specifically, it is suggested that achieving well-being requires an in-depth analysis of the term "quality of life" as an indicator of the urban strategy that is strictly connected with the use of green, with strategies for intensification, and with a sensation of safety in public space.

In all cases, there is an indisputable link between well-being and spatiality. The European cities presented here have selected strategies that are closely linked to urban development and encourage urban innovation environment, whether with the aim of digital technologies in the case of Barcelona and Vienna, or the integration of innovative spatial features (living lab) in the case of Copenhagen. In that sense, the authors argue that the majority of these initiatives focus on the spatial aspect by reinforcing the regeneration of the urban fabric rather than

TABLE 1 Highlighting the four pillars in the work of smart city strategies.

City well-being in smart city strategies focused in the public space of the city					
City	**Ongoing smart city strategies that improve well-being**	**How they improve well-being (*authors' evaluation*)**	**Urban technologies**	**Methods**	**How they provide feedback toward specific well-being indicators (*authors' evaluation*)**
Barcelona	**"Superblocks"**	• Give more space for walking cycling (*activity*) • Discourage car use (*safety*) • Contribute to lower levels of background noise and pollution (*safety*) • Creation of new green areas (*green areas*) • Creation of new recreation areas (*sociability*) • Encourages participatory process (*sociability*)	• "Sentilo" open-source platform	Open-source software, open network of sensors, big data, crowdsourcing, user/citizen engagement	Collects and depicts information for – Temperature, noise, quality of air (*comfort-safety*) – Flow of people, bicycles, cars (*activity*)
			• "Bicing" app	Big data, user/citizen engagement	Provides information – Flow of bicycles-optimization of vehicle transit (*activity*)
			• "Smart Citizen" open-source platform	Crowdsourcing, geolocation, arduino kits, mobile sensors, user/citizen engagement	Extracts information through – The use of urban sensor kits which upload environmental related values (air, noise, lightning) (*comfort-safety*)
			*Information gathered from, https://www.bicing.barcelona/, https://icos.urenio.org/applications/smart-citizen/, https://energy-cities.eu/best-practice/superblocks-free-up-to-92-of-public-space-in-barcelona/, https://icos.urenio.org/applications/sentilo/		

Copenhagen	**"Street Lab"** Urban laboratory (living lab)	Measures and establishes databases for – Air quality (*safety*) – Care of urban nature (*green areas*)	• "PLUSnet system"	ITS (IntelligentTrafficSystems), GPS	Through ITS – Traffic lights dynamically transform the streets and prioritize biking routes "green wave" (*comfort-activity*)
	"Urban Life Accounts" (as supportive action for the "Co-Create Copenhagen vision for 2025")	Measures, maps and annually monitors the status of everyday urban life trends (*sociability, activity, safety*)			Provides information about – Green routes, bicycle superhighways shortcuts (*safety, sociability*)
			*Information gathered from https://urbandevelopmentcph.kk.dk/artikel/copenhagen-solutions-lab, https://kk.sites.itera.dk/apps/kk_pub2/index.asp?mode=detalje&id=1534, https://www.centreforpublicimpact.org/case-study/cycle-infrastructure-copenhagen/, https://handshakecycling.eu/resources/copenhagen-bicycle-statement-2018		
Vienna	**"Greening the city with plant tubes"**	• Prevents urban heat islands and improves the microclimate and air quality (*comfort-safety*) • Creation of new green views (*green areas*)	• "Wien gibt Raum" (or "Sharing Public Spaces") digital service	Geodata, GIS, mobile surveying (mobile mapping and imaging technologies)	Captures and maps information about public space and urban furniture and provides an organizational framework for the management of objects, activities and traffic flows in public space (*sociability and activity*)
			• "Smart traffic lights"	Recognition algorithms, weather and environmental data sensors (big data analytics), GPS systems	Extracts information through – Movement patterns (*comfort-safety*) – Temperature, humidity, pollutants (*safety*) Interconnects traffic lights and integrates data in navigation systems to optimize traffic flow and reduce emissions (*comfort-safety*)
			*Information gathered from https://smartcity.wien.gv.at/site/en/approach/framework-strategy/, https://smartcity.wien.gv.at/site/en/wien-gibt-raum/		

Based on authors' elaboration.

merely promoting emerging digital technologies such as sensors and widgets. This aspect relates to the first approach of the intersection of smart and sustainable, in the first part of this chapter: the one that encompasses smart and sustainable, with the second being the target and the first playing the role of the instrument. In the case of the three European cities, smart seems to be the instrument toward the achievement of social and environmental sustainability.

Another finding in the three European cities is the vast amount of data extracted by low-, mid-, and high-digital technological tools, such as platforms, sensors, and data clouds. With the new technological tools, new possibilities are created in terms of creating, managing, and disseminating information to enhance sustainable economic development, resilience, and high quality of life. At the same time, the cases provide the possibility of transforming the existing urban space into a more dynamic space that can respond better to the needs of citizens. Mobility patterns, environmental parameters, and occupancy data are the main bits of information synthesizing the real-time functioning of the city. However, it is observed that while technology can be seen as the enabler and facilitator of creating and sustaining the urban datasets, there is a missing link between the data and the indicators of well-being. The fact that a vast percentage of urban actions can be traced and measured opens up new possibilities toward the analysis of well-being in correlation with the strategic aims of the smart city, without, however, building the specific indicators.

Lastly, it is crucial to highlight that all smart city strategies encompass urban projects as the starting point of a process of change for the ongoing improvement process of public spaces. These projects are tested in a pilot site on a local scale before being implemented into larger areas of the city. These local pilot projects, such as the "Superblocks," act as precursors to urban innovation in a twofold way: on the one hand, they serve as a test bed for mobility patterns and urban planning alternative strategies; on the other hand, they enable a process of social acceptance and urban dialogue. All projects highlight the potentiality of multiple uses of public space and emphasize the significance of citizen engagement and participation, thus promoting broader public engagement (collective intelligence) and involving all stakeholders of the city's innovation ecosystem.

Specifically, in Barcelona, the key innovator for creating public benefits is the dissemination of technologies to both the state and the citizens so that they can cooperate to set their own priorities. In this way, the data and infrastructure that often belong to the property of large companies now belong to everybody. On the other hand, Copenhagen looks at technology as a means to promote quality of life. Consequently, smart city IT solutions will make the city cleaner and healthier and make it work even better. Lastly, in the Smart City Wien Framework, the integration of the social component into all target areas is a crucial element. Economic objectives and environment-related issues are of equal importance with the improvement of citizens' quality of life. To maintain and further promote this high quality of life, the framework constantly highlights the need for the development of new and innovative solutions.

All these efforts reveal the focus of the case studies toward methods that bridge smart city strategies with sustainable futures, therefore building the ground for working toward the SDG 11: "Make cities and human settlements inclusive, safe, resilient, and sustainable." In addition to everyday life, in these cases it emerges that cities need to plan their development in a sustainable and participatory way, protecting cultural and natural heritage, ensuring metropolitan and territorial integrity, and providing comprehensive and resilient responses to the risks posed by change.

Lastly, the authors argue that there is a solid link between spatiality, well-being, and smart city strategies that need to be further investigated. By disentangling this—so far abstract—intersection, city planners could "zoom in" and identify the indicators that link stress and well-being factors with the already shaped urban database of real-time information. Toward this goal, the authors suggest rereading the city to decode the well-being factors and to visualize the subjective information. In this way, a new urban heat map based on a new, subjective, and clean database containing a new level of artificial intelligence could be produced.

References

Ahvenniemi, H., Huovila, A., Pinto-Seppä, I., & Airaksinen, M. (2017). What are the differences between sustainable and smart cities? *Cities*, *60*, 234–245. https://doi.org/10.1016/j.cities.2016.09.009.

Alohan, D., & Calvo, M. (2020). COVID-19 outbreaks at correctional facilities demand a health equity approach to criminal justice reform. *Journal of Urban Health*, *97*, 342–347. https://doi.org/10.1007/s11524-020-00459-1.

Angelidou, M. (2016). Four European smart city strategies. *International Journal of Social Science Studies*, *4*(4). https://doi.org/10.11114/ijsss.v4i4.1364.

Angelidou, M., Psaltoglou, A., Komninos, N., Kakderi, C., Tsarchopoulos, P., & Panori, A. (2017). Enhancing sustainable urban development through smart city applications. *Journal of Science and Technology Policy Management*, *9*(2). https://doi.org/10.1108/JSTPM-05-2017-0016.

Bassi, A. (2017). Looking at smart cities with an historical perspective. In V. Angelakis, E. Tragos, H. C. Pöhls, A. Kapovits, & A. Bassi (Eds.), *Designing, developing, and facilitating smart cities: Urban design to IoT solutions*. Switzerland: Springer International Publishing.

Bibri, S. E. (2019). On the sustainability of smart and smarter cities in the era of big data: An interdisciplinary and transdisciplinary literature review. *Journal of Big Data*, *6*(25). https://doi.org/10.1186/s40537-019-0182-7.

Bibri, S. E., & Krogstie, J. (2017). Smart sustainable cities of the future: An extensive interdisciplinary literature review. *Sustainable Cities and Society*, *31*, 183–212. https://doi.org/10.1016/j.scs.2017.02.016.

Bifulco, F., Tregua, M., Amitrano, C. C., & D'Auria, A. (2016). ICT and sustainability in smart cities management. *International Journal of Public Sector Management*, *29*(2), 132–147. https://doi.org/10.1108/IJPSM-07-2015-0132.

Bouskela, M., Casseb, M., Bassi, S., De Luca, C., & Facchina, M. (2016). *The road toward Smart Cities: Migrating from traditional city management to the smart city*. IDB Library Publisher. 15/07/2020. Retrieved from https://publications.iadb.org/publications/english/document/The-Road-toward-Smart-Cities-Migrating-from-Traditional-City-Management-to-the-Smart-City.pdf.

Capolongo, S., Rebecchi, A., Dettori, M., Appolloni, L., Azara, A., Buffoli, M., et al. (2018). Healthy design and urban planning strategies, actions, and policy to achieve salutogenic cities. *International Journal of Environmental Research and Public Health, 15*(12), 2698. https://doi.org/10.3390/ijerph15122698.

Carr, S., Francis, M., Rivlin, L. G., & Stone, A. M. (1993). *Public space, Cambridge series in environment and behavior.* New York, NY: Cambridge University Press.

Centre for Public Impact. (2016a). *Cycle infrastructure in Copenhagen.* 10/07/2020. Retrieved from https://www.centreforpublicimpact.org/case-study/cycle-infrastructure-copenhagen/.

Centre for Public Impact. (2016b). *Green waves for bicycles in Copenhagen.* 17/07/2020. Retrieved from https://www.centreforpublicimpact.org/case-study/green-waves-bicycles-copenhagen/#measurement.

Centre for Urban Design and Mental Health (UD/MH). (2016). *Mind the GAPS framework: The impact of urban design and mental health and wellbeing.* 01/07/2020. Retrieved from https://www.urbandesignmentalhealth.com/mind-the-gaps-framework.html.

Coldwell, D. F., & Evans, K. L. (2018). Visits to urban green-space and the countryside associate with different components of mental well-being and are better predictors than perceived or actual local urbanisation intensity. *Landscape and Urban Planning, 175*, 114–122. https://doi.org/10.1016/j.landurbplan.2018.02.007.

Corburn, J., Vlahov, D., Mberu, B., Riley, L., Caiaffa, W. T., Rashid, S. F., et al. (2020). Slum health: Arresting COVID-19 and improving well-being in urban informal settlements. *Journal of Urban Health, 97*, 348–357. https://doi.org/10.1007/s11524-020-00438-6.

Cucuzzella, C., & Goubran, S. (2019). Infrastructure as a deeply integrated sustainable urban project. *Journal of Sustainability Research, 1*(1). https://doi.org/10.20900/jsr20190005.

De Falco, S. (2019). Are smart cities global cities? A European perspective. *European Planning Studies, 27*(4), 759–783. https://doi.org/10.1080/09654313.2019.1568396.

De Silva, M. J., McKenzie, K., Harpham, T., & Huttly, S. R. (2005). Social capital and mental illness: A systematic review. *Journal of Epidemiology and Community Health, 59*(8), 619–627. https://doi.org/10.1136/jech.2004.029678.

Evans, G. W. (2003). The built environment and mental health. *Journal of Urban Health, 80*(4), 536–555. https://doi.org/10.1093/jurban/jtg063.

Evans, G. W., & Cohen, S. (1987). *Handbook of environmental psychology. Vol. 1* (pp. 571–610). New York: Wiley.

Fraga, J. (2019). *Here's how living in a city can mess with your mental health.* 16/07/2020. Retrieved from https://www.healthline.com/health/mental-health/living-in-a-city#1.

Francis, J., Wood, L. J., Knuiman, M., & Giles-Corti, B. (2012). Quality or quantity? Exploring the relationship between Public Open Space attributes and mental health in Perth, Western Australia. *Social Science & Medicine, 74*(10), 1570–1577. https://doi.org/10.1016/j.socscimed.2012.01.032.

Guedes, A. L. A., Alvarenga, J. C., Goulart, M. S. S., Rodriguez, M. V., & Soares, C. A. P. (2018). Smart cities: The main drivers for increasing the intelligence of cities. *Sustainability, 10*(9), 3121. https://doi.org/10.3390/su10093121.

Haarstad, H., & Wathne, M. W. (2019). Are smart city projects catalyzing urban energy sustainability? *Energy Policy, 129*, 918–925. https://doi.org/10.1016/j.enpol.2019.03.001.

Hajrasouliha, A., Rio, V., Francis, J. L., & Edmonson, J. (2018). Urban form and mental wellbeing: Scoping a theoretical framework for action. *Journal of Urban Design and Mental Health, 5*(10). 17/07/2020. Retrieved from https://www.urbandesignmentalhealth.com/journal-5-urban-form-and-mental-wellbeing.html.

Herzog, T. R., Maguire, P., & Nebel, M. B. (2003). Assessing the restorative components of environments. *Journal of Environmental Psychology, 23*(2), 159–170. https://doi.org/10.1016/S0272-4944(02)00113-5.

Hidalgo, M. C., Berto, R., Galindo, M. P., & Getrevi, A. (2006). Identifying attractive and unattractive urban places: Categories, restorativeness and aesthetic attributes. *MedioAmbiente y ComportamientoHumano*, *7*(2), 115–133. 17/07/2020. Retrieved from https://www.researchgate.net/publication/260058996_Identifying_attractive_and_unattractive_urban_places_Categories_restorativeness_and_aesthetic_attributes.

Hunter, R. F., Cleland, C., Cleary, A., Droomers, M., Wheeler, B. W., Sinnett, D., et al. (2019). Environmental, health, wellbeing, social and equity effects of urban green space interventions: A meta-narrative evidence synthesis. *Environment International*, *130*. https://doi.org/10.1016/j.envint.2019.104923.

Komninos, N. (2016). Afterword: Smart urban futures. In T. Yigitcanlar (Ed.), *Technology and the city: Systems, applications and implications* (pp. 172–181). London: Routledge. https://doi.org/10.4324/9781315739090.

Lin, C., Zhao, G., Yu, C., & Wu, Y. J. (2019). Smart city development and residents' well-being. *Sustainability*, *11*(3), 676. https://doi.org/10.3390/su11030676.

Litman, T., & Victoria Transport Policy Institute. (2020). *Urban sanity, understanding urban mental health impacts and how to create saner, happier cities*. 17/07/2020. Retrieved from https://www.vtpi.org/urban-sanity.pdf.

Maller, C., Townsend, M., Pryor, A., Brown, P., & St Leger, L. (2006). Healthy nature healthy people: 'Contact with nature' as an upstream health promotion intervention for populations. *Health Promotion International*, *21*(1), 45–54. https://doi.org/10.1093/heapro/dai032.

McCay, L. (2017). Designing mental health into cities. *Urban Design Group Journal*, *142*, 25–27. 17/07/2020. Retrieved from http://www.udg.org.uk/sites/default/files/publications/UD142_magazine%20Health%20and%20Urban%20Design.pdf.

Ompad, D. S., Galea, S., & Vlahov, D. (2017). Urban health systems: Overview. In *International encyclopedia of public health* (2nd ed., pp. 311–317). https://doi.org/10.1016/B978-0-12-803678-5.00481-1.

Pineda, V. S., & Corburn, J. (2020). Disability, urban health equity, and the coronavirus pandemic: Promoting cities for all. *Journal of Urban Health*, *97*, 336–341. https://doi.org/10.1007/s11524-020-00437-7.

Pineo, H., Glonti, K., Rutter, H., Zimmermann, N., Wilkinson, P., & Davies, M. (2020). Use of urban health indicator tools by built environment policy- and decision-makers: A systematic review and narrative synthesis. *Journal of Urban Health*, *97*, 418–435. https://doi.org/10.1007/s11524-019-00378-w.

The City of Copenhagen. (2012). *City of Copenhagen's bicycle strategy 2011-2025*. 12/07/2020. Retrieved from https://handshakecycling.eu/resources/city-copenhagen%E2%80%99s-bicycle-strategy-2011-2025.

The City of Copenhagen. (2015a). *Copenhagen Solutions Lab*. 18/07/2020. Retrieved from https://urbandevelopmentcph.kk.dk/artikel/copenhagen-solutions-lab.

The City of Copenhagen. (2015b). *Co-create Copenhagen*. 18/07/2020. Retrieved from https://kk.sites.itera.dk/apps/kk_pub2/index.asp?mode=detalje&id=1534.

The City of Copenhagen. (2019). *Smart city*. 18/07/2020. Retrieved from https://urbandevelopmentcph.kk.dk/indhold/smart-city.

Thomas, F., Sabel, C. E., Morton, K., Hiscock, R., & Depledge, M. H. (2014). Extended impacts of climate change on health and wellbeing. *Environmental Science & Policy*, *44*, 271–278. https://doi.org/10.1016/j.envsci.2014.08.011.

Trencher, G., & Karvonen, A. (2019). Stretching "smart": Advancing health and well-being through the smart city agenda. *Local Environment*, *24*(7), 610–627. https://doi.org/10.1080/13549839.2017.1360264.

UN-Habitat. (2019). *Annual progress report*. 15/07/2020. Retrieved from https://unhabitat.org/sites/default/files/2020/03/annual_report_2019_03022020_posted.pdf.

UN-Habitat & World Health Organization. (2020). *Integrating health in urban and territorial planning: A sourcebook*. 18/07/2020. Retrieved from https://apps.who.int/iris/handle/10665/331678.

United Nations. (2015). *Transforming our world: The 2030 agenda for sustainable development*. 17/07/2020. Retrieved from https://sustainabledevelopment.un.org/post2015/transformingourworld.

United Nations. (2017). *UNDP support to the implementation of sustainable development goal three, ensure healthy lives and promote well-being for all at all ages*. 10/07/2020. Retrieved from https://www.researchgate.net/publication/283451447_Goal_3_Ensure_healthy_lives_and_promote_well-being_for_all_at_all_ages.

United Nations. (2019). *Big data and the 2030 agenda for sustainable development*. 14/07/2020. Retrieved from https://www.unescap.org/sites/default/files/Final%20Draft_%20stock-taking%20report_For%20Comment_301115.pdf.

Vienna Municipal Administration. (2019a). *Smart City Wien framework strategy 2019–2050, Vienna's strategy for sustainable development*. 12/07/2020. Retrieved from https://smartcity.wien.gv.at/site/en/approach/framework-strategy/.

Vienna Municipal Administration. (2019b). *Sharing public spaces in Vienna*. 17/07/2020. Retrieved from https://smartcity.wien.gv.at/site/en/wien-gibt-raum/.

Visvizi, A., & Lytras, M. D. (2018a). It's not a fad: Smart cities and smart villages research in European and global contexts. *Sustainability*, *2018*(10), 2727. https://doi.org/10.3390/su10082727.

Visvizi, A., & Lytras, M. (2018b). Rescaling and refocusing smart cities research: From mega cities to smart villages. *Journal of Science and Technology Policy Management*. https://doi.org/10.1108/JSTPM-02-2018-0020.

Visvizi, A., & Lytras, M. D. (2019a). Sustainable smart cities and smart villages research: Rethinking security, safety, well-being, and happiness. *Sustainability*, *12*(1), 1–4. https://doi.org/10.3390/su12010215.

Visvizi, A., & Lytras, M. D. (2019b). Smart cities research and debate: What is in there? In A. Visvizi, & M. D. Lytras (Eds.), *Smart cities: Issues and challenges: Mapping political, social and economic risks and threats* Elsevier, ISBN:9780128166390. https://www.elsevier.com/books/smart-cities-issues-and-challenges/lytras/978-0-12-816639-0.

Vujčić, M., Tomicevic-Dubljevic, J., Zivojinovic, I., & Toskovic, O. (2019). Connection between urban green areas and visitors' physical and mental well-being. *Urban Forestry & Urban Greening*, *40*, 299–307. https://doi.org/10.1016/j.ufug.2018.01.028.

Wood, L., Hooper, P., Foster, S., & Bull, F. (2017). Public green spaces and positive mental health—Investigating the relationship between access, quantity and types of parks and mental wellbeing. *Health & Place*, *48*, 63–71. https://doi.org/10.1016/j.healthplace.2017.09.002.

World Health Organization. (2019). *Evaluating the WHO Assessment Instrument for Mental Health Systems by comparing mental health policies in four countries*. 15/07/2020. Retrieved from https://www.who.int/bulletin/volumes/86/6/07-042788/en/.

Zaman, A. U. (2014). Roadmap towards zero waste cities. *International Journal of Waste Resources*, *4*(2), 1000e1106. https://doi.org/10.4172/2252-5211.1000e106.

Further reading

Copenhagen Bicycle Statement 2018, 17/07/2020, Retrieved from: https://handshakecycling.eu/resources/copenhagen-bicycle-statement-2018.???Mou-te de manera sostenible per Barcelona!,

17/07/2020. Retrieved from: https://www.bicing.barcelona/.Mou-te de manera sostenible per Barcelona!, 17/07/2020. Retrieved from: https://www.bicing.barcelona/.

Sentilo, An open source sensor and actuator platform designed to be used by any city, 17/07/2020. Retrieved from: https://icos.urenio.org/applications/sentilo/.

Smart Citizen, An open source platform for crowdsourced environmental monitoring, 17/07/2020, Retrieved from: https://icos.urenio.org/applications/smart-citizen/.

"Superblocks" free up to 92% of public space in Barcelona!, 17/07/2020, Retrieved from: https://energy-cities.eu/best-practice/superblocks-free-up-to-92-of-public-space-in-barcelona/.

Chapter 11

Recognizing intra-urban disparities in smart cities: An example from Poland

Piotr Maleszyk
Maria Curie-Skłodowska University, Lublin, Poland

1. Introduction

Contemporary cities have become complex and multifaceted organisms coping with various economic, social, and environmental challenges. These challenges will certainly continue to increase, as will the total urban population: 55% of the world's population lives in urban areas, while 68% is projected to be urban by 2050 (UN, 2018). These developments are accompanied by the decline of industrial cities, such as Detroit, Bochum, or Belfast, that have experienced difficult transitions toward services and a knowledge-based economy. Nevertheless, the global role of cities and urban areas continues to expand, as does the role of city governments (Barber, 2013; Visvizi & Lytras, 2019a, 2019b). The increasing pace of urbanization, along with the growing role of cities in world politics and the global economy, has resulted in the inclusion of urban areas in the 2030 Agenda for Sustainable Development (Kosovac, Acuto, & Jones, 2020). The role of cities in achieving sustainable development has been explicitly stated in the 11th Sustainable Development Goal (SDG 11) calling for making cities inclusive, safe, resilient, and sustainable.

Successful cities need to produce wealth and innovation, increase the well-being of their residents, protect their environment, integrate a population from diverse backgrounds, and increase civic activity. They must attract talent, increase the resilience of their urban areas to various shocks, and successfully cooperate with other cities in global networks. At the same time, residents' knowledge and expectations are also growing, and with them comes increased pressure for effective urban governance (Lytras, Visvizi, Chopdar, Sarirete, & Alhalabi, 2021; Lytras, Visvizi, & Sarirete, 2019). Thus, policies and tools that aim to foster the governance of urban areas should offer solutions supporting urban authorities in meeting the abovementioned goals and challenges.

Smart Cities and the UN SDGs. https://doi.org/10.1016/B978-0-323-85151-0.00011-7

The concept of “smart city” remains vague with no commonly accepted definition, used in many ways that are not always consistent (Visvizi & Lytras, 2018). Not only does a plethora of definitions exist, but there is also no one-size-fits-all definition (Albino, Berardi, & Dangelico, 2015; Anthopoulos, Janssen, & Weerakkody, 2019; Manville et al., 2014; Visvizi & Lytras, 2019a). This conceptual disarray partly stems from the complexity of the urban system and the interdependencies among its elements. Urban authorities’ desire to call their cities “smart” further contributes to expanding and blurring the scope of this concept. Nevertheless, a sine qua non of smart city initiatives is the usage of new technologies, though not as a goal in itself, but a means of significantly enhancing the performance or quality of a city, ultimately bolstering its citizens’ engagement (Borsekova & Nijkamp, 2018; Mora, Bolici, & Deakin, 2017). Further, although the debate on where the notion of a smart city begins and where it ends remains unresolved, smart governance is broadly accepted as one of its indispensable dimensions (Caragliu, Del Bo, & Nijkamp, 2011; Giffinger et al., 2007; Neirotti, De Marco, Cagliano, Mangano, & Scorrano, 2014; Ruhlandt, 2018).

In general, the literature on smart governance explores how the application of information and communication technology (ICT) reinforces citizens’ participation and strengthens the capacity of urban authorities and stakeholders to tackle a wide variety of problems (Borsekova & Nijkamp, 2018; Ruhlandt, 2018). The notion of “smart governance” is related to the preexisting concepts of e-administration (Meijer & Bolívar, 2016), Government 2.0, and citizen-sourcing (Nam, 2012). The specialized literature on the subject links smart city governance and its outcomes to the appealing approach of urban sustainability (Ahvenniemi, Huovila, Pinto-Seppä, & Airaksinen, 2017; Tomor, Meijer, Michels, & Geertman, 2019; Visvizi & Lytras, 2019b). In this chapter, we follow this approach while adopting Tomor et al. (2019, p. 3)’s definition of smart governance—“technology-enabled collaboration between citizens and local governments to advance sustainable development.”

Given the conceptual fuzziness of the smart city, deciding whether a particular urban innovation fits into the scope of smart city governance might not be very clear-cut. In their extensive literature review, Meijer and Bolívar (2016) clarify this issue by identifying four levels of conceptualization of smart city governance, which differ in terms of the degree of transformation of government structures needed to make cities smarter: (1) governance of a smart city, (2) smart decision-making, (3) smart administration, and (4) smart urban collaboration. The first level entails no need for governance structures and processes to transform. The second is not about restructuring an organization or government institution, but it does emphasize the need to restructure decision-making by using network technologies. Smart administration is regarded as a new form of electronic governance that uses sophisticated information technologies to interconnect and integrate information, processes, institutions, and physical infrastructure to better serve citizens and communities. This approach may be considered a redesign of the idea of e-government, which has already been present in the literature for several decades. The fourth and most transformative level of smart governance requires not only the transformation of government institutions, but also of a wide range of urban stakeholders. It emphasizes the widespread adoption of a community-based, citizen-centric model of governance, with greater connectivity facilitated by new technologies. Naturally, smart urban collaboration projects are viewed as most valuable.

The implementation of smart administration and smart urban collaboration in a manner that could address pressing urban issues entails the need to gather new data from city dwellers and further utilize the knowledge gathered in a process of co-creation. With this context in mind, this chapter aims to present and discuss the application of an online tool for recognizing satisfaction with urban public amenities in neighborhood areas that has been implemented in the city of Lublin (Poland). The distinguishing feature of this tool is that a large number of credible assessments enable us to recognize the perception of intra-urban differences among residents—an issue that is difficult to address empirically and thus rarely explored, yet critical in efforts to make urban areas more inclusive and sustainable in the context of SDG 11. The chapter further discusses the potential of such tools to strengthen the rationality of decisions adopted by urban authorities and to guide the urban institutional framework toward a community-based model of governance.

The remaining part of the chapter is structured as follows. The first section sets the scene by summarizing trends in smart governance in Poland and in Lublin. In the following part, we analyze our dataset using differentiation measures, and aggregate indicators for neighborhood areas are calculated to reveal and discuss the urban geography of neighborhood satisfaction across 27 urban districts in Lublin. The conclusion summarizes the key findings and indicates directions for the future.

2. Trends in smart governance in Poland

Understanding the developments taking place in smart governance in Poland, particularly those entailing smart decision-making and collaboration relating to the urban space, requires a presentation of the historical patterns of spatial development of Polish cities. Beginning in the 1970s, large-scale, prefabricated concrete housing projects (blocks of flats) were built to meet the increasing housing needs, often with poor quality materials. Meanwhile, city centers, including traditional market squares, etc., remained underinvested, underdeveloped, and in essence deserted. After the fall of communism, cities faced continued devastation of urban centers and the degradation of concrete housing estates inhabited largely by industrial workers. Simultaneously, urban areas experienced social tensions that arose after a chain of bankruptcies of state-owned industrial plants and aggressive suburbanization of more affluent residents. These processes often resulted in socio-spatial segregation within urban areas, increasing social inequality and gentrification (cf. Marcińczak, 2012). Although local government had been established, financial constraints restricted many an opportunity to revive cities and improve the quality of life for more than a decade. Only Poland's accession to the European Union and the intake of structural funds allowed investments in infrastructure and the regeneration of urban centers (cf. Visvizi, Matysek-Jedrych, & Mroczek-Dąbrowska, 2020). EU membership also triggered the economic growth in cities and facilitated the transition from local government management to the concept of smart governance.

The implementation of smart urban collaboration in Poland would not have been possible without favorable institutional changes. The fundamental legal change toward decentralization was marked by the creation of self-governing structures in 1990, and further by a 1999 administrative reform on the national level. With regard to informal institutions, the increase in overall civic engagement has also been conducive to modernization toward smart governance, though typical urban citizens still remain less concerned about local affairs than their counterparts in the EU15 countries (Pistelok & Martela, 2019). Nevertheless, since the 2000s, a clear turn toward more informal grassroots activity has been observed, above all in the field of urban activism (Polanska, 2018). This has been accompanied by an increasing trend among urban authorities that consists in offering opportunities for public participation, particularly in regional capital cities (Pistelok & Martela, 2019). However, nondigital, one-time commitments and top-down measures are still prevalent (see, e.g., Gawłowski & Jarosz, 2019). With respect to strategic planning, there is a clear tendency toward increasing the emphasis on "smart city" in national strategic documents such as the National Urban Policy 2023 (NUP 2023, 2015), the Strategy for Responsible Development (SRD, 2017), and the National Regional Development Strategy 2030 (NRDS, 2030, 2019). Nevertheless, the idea of smart governance in these documents is reduced to smart administration, while public participation is not embedded in the "smart city" concept. A more in-depth and comprehensive approach to the smart city and smart governance is presented in local urban strategies that extend until 2030, adopted recently by several large Polish cities (e.g., Warsaw, Wroclaw, Cracow). On the whole, the pursuit of the "smart city" idea is the sum of individual initiatives by leading Polish cities with little national assistance and coordination. Given the central government's priority of supporting small and medium cities rather than regional capitals, which is explicitly declared in NRDS 2030 (2019, pp. 40–41), little central co-funding and coordination is expected in the near future.

The implementation of the smart city concept in Polish cities in its early phase (2008–15) showed a strong focus on investments in such areas as energy, transport systems (or their individual components), and public safety. A more detailed assessment reveals that investments in transport systems indeed entailed many smart and innovative components, whereas most projects in other areas were marked by a poorly developed or even completely missing component that accounted for the use of state-of-the-art ICT. Energy savings were accomplished largely through the thermo-modernization of public buildings or by installing LED street lighting, while public safety was ensured by public CCTV monitoring. Further, smart city initiatives have often been driven by technological firms and implemented without a comprehensive approach (e.g., Sikora-Fernandez, 2017, 2018). These difficulties in implementation are, to some extent, country specific. Decision-making regarding the selection and financing of urban investments in Poland is multileveled: urban municipalities (local level) have been decisive, yet regional authorities have set priorities for spending EU funds, while strategic decisions regarding the allocation of EU funds have been made at the national level. As a result, many innovative local projects were halted, as they were less aligned with the eligibility criteria established earlier on a regional or central level than less innovative infrastructure projects. This contributed to inertia in the implementation of highly innovative smart city projects.

Recent smart city initiatives in Poland are driven more clearly by public expectations and marked by a stronger technological (often IoT and AI) component (Innovation map, 2020), though the number of initiatives that refer to "smart governance" is still lower than in other countries that have invested in smart city development. Nevertheless, some projects have emerged that strive for achieving a better quality of life while enhancing information exchange between urban authorities and city dwellers. Some examples include applications that offer real-time information or notification of failures and outages in public amenities, spatial information systems and open-data platforms, e-consultation platforms, urban labs, and comprehensive e-service systems (Knight Frank, 2018a, 2018b; Innovation map, 2020). Many of these projects are still ongoing.

3. The implementation of smart city concept: The Lublin experience

Lublin is the capital city of the Lubelskie Region, located on the eastern border of the EU. It ranks ninth in population size, with 340,000 residents in the city and approximately 550,000 in the urban area. Lublin appears to be a representative example of a Polish regional city in terms of smart city development, though it excels in the area of smart governance.[a] In 2008–15, a leading large-scale "smart" project in Lublin targeted the development of a sustainable transport system and included, among others, freeway management, transit management, real-time public transport information, the purchase of zero-emission buses, electronic payments, and reforms in pricing. In addition, an urban bicycle-sharing scheme was introduced and developed in this period. The city has pursued its environmental goals largely by installing LED street lighting and through thermomodernizing public buildings. Lublin was among the first cities to implement an automated system for waste collection and

a. See: https://www.beesmart.city/cities/lublin-PL-EU-f57190 for a description of the projects and cross-national benchmarks.

container monitoring using RFID technology (2013). This period witnessed the development of free wi-fi hot spots, public CCTV monitoring, and the introduction of a local geoportal that integrates spatial data. Finally, an urban data center was established to provide the core infrastructure for further ICT-heavy urban projects. Altogether, the abovementioned trends accurately represent the general tendencies observed in large Polish cities. With respect to projects that entail smart governance and urban collaboration, one might find examples of conventional initiatives implemented in other Polish cities, such as participatory budgeting with online voting options or efforts to increase the number of partnerships and projects between the public sector and NGOs, as well as more innovative and ICT-based projects, such as a consortium of seven academic and cultural libraries that creates and publishes digital collections, or *fix.it*—an app that allows users to report and track failures and outages in public amenities and utilities in real time. Lublin also has also built a robust business ecosystem through the Lublin IT Upland project (since 2011), implementing the triple-helix model of collaboration between public sector actors, business, academia, and NGOs (Maleszyk & Sagan, 2018) in one of the earliest applications of the RIS3 concept (cf. Foray, David, & Hall, 2009) in urban areas in CEE countries.

As a result of the 2014–20 EU Multiannual Financial Framework (MFF), the period after 2015 was characterized by distinctly more innovative smart city projects. Given the scope of this chapter, actions that applied the smart governance idea in some way are especially noteworthy. The most relevant project in Lublin aims at establishing and developing an e-services platform, which entails the development of a geoportal with an analytical module and a 3D/4D city model, a new open-data platform, an e-participation panel, and an e-contact center. Further, Lublin has already implemented a comprehensive IT system for educational institutions, which improves these institutions' management and offers a broad range of e-services dedicated to teachers, pupils, and their parents. Alongside this, "Foresight Lublin 2050" was an intangible project that mobilized various types of stakeholders to explore Lublin's identity and design future scenarios for the city. Lublin also obtained an ISO 37120 Certificate, which evaluates cities on more than 100 indicators that measure the efficiency of public services and the quality of life in line with the concepts of a "sustainable city" and "smart governance." Stakeholder collaboration has been enhanced through the introduction of the Green Participatory Budget (in 2017), the Youth Participatory Budget (in 2019), and, most recently, through the launch of a process to build a distinct Lublin 2030 Strategy, with a strong focus on citizen participation and collaboration with various stakeholders. The popularity of participatory budget voting as well as the increasing activity of district councils both encouraged the city to introduce an online tool aiming to collect residents' assessments regarding the accessibility of public urban amenities and their quality in the city's neighborhoods. The central idea of this research was to recognize intra-urban differences, which allowed for a more locally tailored approach in addressing urban inequalities, more efficient investment planning—and subsequently more sustainable spatial development of urban areas.

4. Recognizing patterns in intra-urban disparities in neighborhood satisfaction

4.1 Research design and data

An online survey was conducted from September 23 to October 10, 2018, along with online voting on participatory budget projects. All voters were asked to fill out a short questionnaire in which they were asked to assess, among others, the quality of public urban amenities in their districts on an interval scale ranging from 1 (very bad) to 7 (very good), with an additional "I don't know/not applicable" nonresponse item. The categories of public amenities included in the scope of the assessment were as follows: green areas, accessibility by car, street condition, pavement condition, availability of parking space, city cleanliness, accessibility by public transport, sport facilities and amenities, kindness of local residents, adaptation of public spaces for persons with disabilities, cycling lanes, availability of kindergartens, quality of education in primary schools, and children's playgrounds. Additionally, participants were asked to assess the safety (separately during the day and at night) and air quality of the neighborhood as well as the kindness of local residents—three space-related categories that do not fall into the "public amenities" category, yet apply to the concept of "public goods" and significantly influence the well-being of urban residents.

The survey was completed by 5753 residents aged 15 and over, out of 19,809 voters in the same age group, which translates into a response rate of 32%—a very high percentage with respect to most online surveys. The numbers of participants exceeded 50 in 21 out of 27 districts, surpassed 100 in 18 districts. Overall, the participants accounted for 2% of the total population aged 15 and over. A summary of respondent statistics is presented in Table 1.

One of the key issues with government-citizen communication enhanced by ICT is its nonrepresentativeness (Nam, 2012). In extreme cases, urban citizens are represented by a small group of urban activists willing to undertake time-consuming actions and whose ideas are often skewed and inadequately reflect the opinions and needs of the "silent majority" of urban residents. The common reality of public participation is that those who show up in open meetings either have negative views or are passionate about particular issues or values, while less assertive people are severely underrepresented (Haklay, Jankowski & Zwoliński, 2018). This also applies to cities in Poland (Pistelok & Martela, 2019), including Lublin. In this regard, many appealing high-tech solutions that allow for real-time collaboration, often labeled as "smart," might actually fail to increase the number and representativeness of urban residents involved in citizen participation. In turn, recognizing intra-urban differences in neighborhood satisfaction requires maximizing the number of individuals expressing their opinion. In this context, providing a simple, intuitive, and inclusive tool was the optimal framework for our research.

Our questionnaire accompanied Lublin's participatory budget voting, which is a highly popular citizen participation measure in Lublin, engaging on average 10%–15% of the city's residents each year. Although not fully representative,

TABLE 1 Summary statistics.

	Respondents Number	%	City population (aged 15 and above) %
Age			
15–24	496	8.6	10.3
25–34	1943	33.8	17.7
35–44	2136	37.1	19.2
45–54	678	11.8	14.2
55–64	327	5.7	15.7
65 and above	173	3.0	22.9
Sex			
Male	2512	43.7	45.3
Female	3197	55.6	54.7
No answer	44	0.8	–
District			
Abramowice (1)	20	0.3	0.6[a]
Bronowice (2)	419	7.3	4.5[a]
Czechów Południowy (3)	273	4.7	7.5[a]
Czechów Północny (4)	310	5.4	5.2[a]
Czuby Południowe (5)	589	10.2	5.3[a]
Czuby Północne (6)	543	9.4	8.5[a]
Dziesiąta (7)	453	7.9	6.8[a]
Felin (8)	168	2.9	2.2[a]
Głusk (9)	39	0.7	0.6[a]
Hajdów-Zadębie (10)	30	0.5	0.7[a]
Kalinowszczyzna (11)	357	6.2	7.3[a]
Konstantynów (12)	79	1.4	2.5[a]
Kośminek (13)	126	2.2	4.0[a]
Ponikwoda (14)	326	5.7	3.9[a]
Rury (15)	382	6.6	9.4[a]

TABLE 1 Summary statistics—cont'd

	Respondents Number	%	City population (aged 15 and above) %
Sławin (16)	245	4.3	3.4[a]
Sławinek (17)	156	2.7	2.2[a]
Stare Miasto (18)	16	0.3	0.7[a]
Szerokie (19)	98	1.7	1.0[a]
Śródmieście (20)	164	2.9	6.1[a]
Tatary (21)	144	2.5	3.6[a]
Węglin Południowy (22)	316	5.5	2.6[a]
Węglin Północny (23)	99	1.7	1.2[a]
Wieniawa (24)	149	2.6	3.8[a]
Wrotków (25)	186	3.2	4.5[a]
Za Cukrownią (26)	33	0.6	1.1[a]
Zemborzyce (27)	33	0.6	0.9[a]
Total	5753	100	100

[a]*Estimates based on data on registered city population aged 18 and above.*

it attracts a large pool of residents with above-average interest in various neighborhood-related matters. The involvement of individuals with high social capital is favorable for the reliability of the data obtained. The sample accurately reflects the gender distribution of the urban population. Furthermore, the survey ensured sufficient representation from all 27 districts, while differences in the absolute number of participants across districts derive largely from population differences rather than survey bias (see Table 1). In terms of age, groups aged 55 and above are underrepresented, as older people are less likely to vote via the Internet. Nevertheless, unlike many other internet tools (see, e.g., Czepkiewicz, Jankowski, & Młodkowski, 2017; Haklay et al., 2018), the middle-aged population was willing to fill out the questionnaire. One might consider applying correction techniques, yet we did not employ a weighting adjustment for two reasons. Firstly, detailed data on citizens' age across districts come from the city registry and thus do not include a large number of unregistered residents who also voted on the participatory budget and filled out the online questionnaire, which could be a further source of bias. Secondly, assessments of local amenities are, according to this dataset, distinguished largely by the quality of the area rather than participants' age.

Further analysis followed a two-step approach. First, we explore differences in neighborhood satisfaction with each type of local urban amenities and public goods across 27 districts by using classical position measures calculated as the ratio of the standard deviation to the mean. We use position measures as data for each district or assessed category show a skewed distribution. Afterward, we calculate aggregate indicators of neighborhood satisfaction across districts in Lublin. We present results obtained by applying equal weighting to the assessed categories, as using an uneven weighting strategy based on public opinion polling has yielded almost identical district hierarchies and distances between assessed units (Maleszyk, 2019).

4.2 Results and discussion

The general impression of respondents' satisfaction with various local public amenities is that intra-urban differences are not very large. Classical coefficients of variation across 27 districts range from 6.1% for safety during the day to 21.8% for availability of parking space. Coefficients of variation calculated from the positional measures (quartile deviation by median) are even lower and range from 3.0% to 16.4%. Detailed position measures for each category are plotted on a box and whisker graph (Fig. 1). The longer the boxes and whiskers are, the greater the differences among the districts for each category.

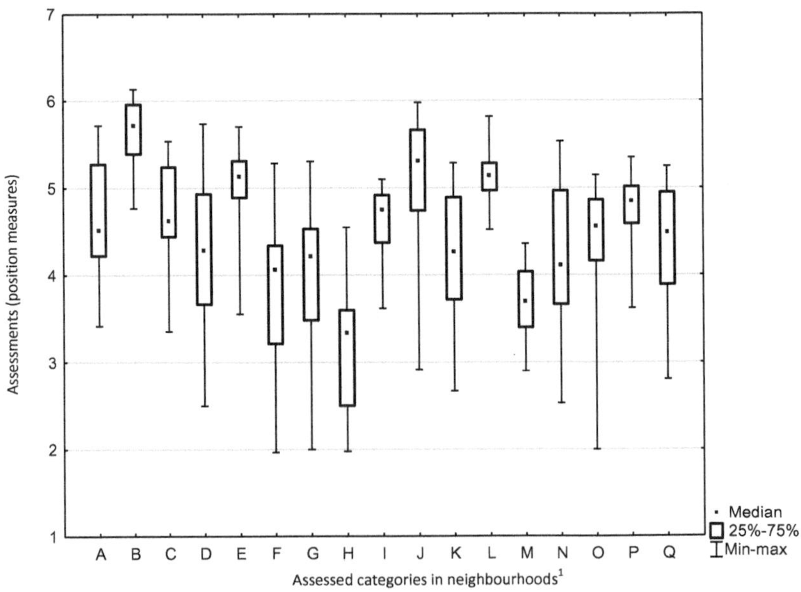

[1]Notes: A: green areas; B: safety in the daytime; C: safety at night; D: air quality; E: accessibility by car; F: street condition; G: pavement condition; H: availability of parking space; I: city cleanliness; J: accessibility by public transport; K: sport s facilities and amenities; L: kindness of local residents; M: adaptation of public spaces for persons with disabilities N: cycling lanes; O: availability of kindergartens; P: quality of education in primary schools; Q: children's playgrounds.

FIG. 1 Differentiation of assessments of local urban amenities across districts.

Although assessments across categories show moderate differentiation, several public amenities are assessed more unevenly than others. This is particularly true for availability of parking space (H), bike paths (N), air quality (D), condition of streets and pavements (F, G), and sport facilities and amenities (K). In contrast, there is actually little difference across districts in terms of kindness of local residents (L), quality of education in primary schools (P), safety in the daytime (B), and city cleanliness (I). Overall, the results indicate that Lublin (and so many other similar regional cities in Poland) is not affected by strong intra-urban inequalities in terms of public goods provision, while the differences that do emerge concern mainly infrastructure for urban mobility.

In the next step, we turn to the analysis of differences across districts to investigate whether high (low) assessments prevail in certain districts. The aggregate indicators—i.e., the unweighted averages of all assessed categories for each of the 27 districts—are depicted in Fig. 2. The main finding is that most districts are similar in terms of the overall indicator of neighborhood satisfaction. The classical coefficient of variation of the aggregate indicators for the 27 districts is barely 8%, which seems low against corresponding measures for specific categories of public amenities. Furthermore, living conditions as perceived by Lublin residents are not strongly differentiated considering that the well-evaluated districts are also the most populous: the 11 leading local areas that scored between 4.6 and 4.92 accounted for 58% of all urban residents, while the two lagging districts with a score below 4.0 were inhabited by 1.2% of Lublin residents.

Although the evidence indicates that intra-urban variation in neighborhood satisfaction is seemingly small, some interesting spatial patterns occur. First, the northwestern part of the city performs better overall than the southeastern districts (see Map 1). This division clearly relates to the main functions of these

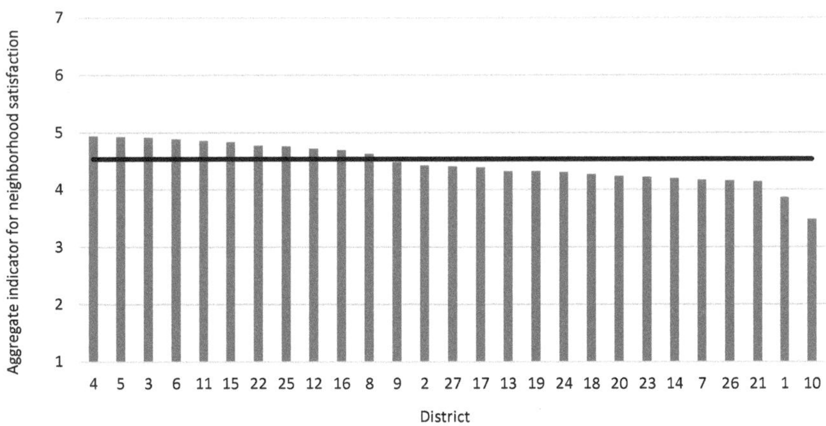

Notes: see table 1 for district names. Horizontal line plots the value of the aggregate indicator for Lublin.

FIG. 2 Aggregate indicators for neighborhood satisfaction in 27 districts in Lublin.

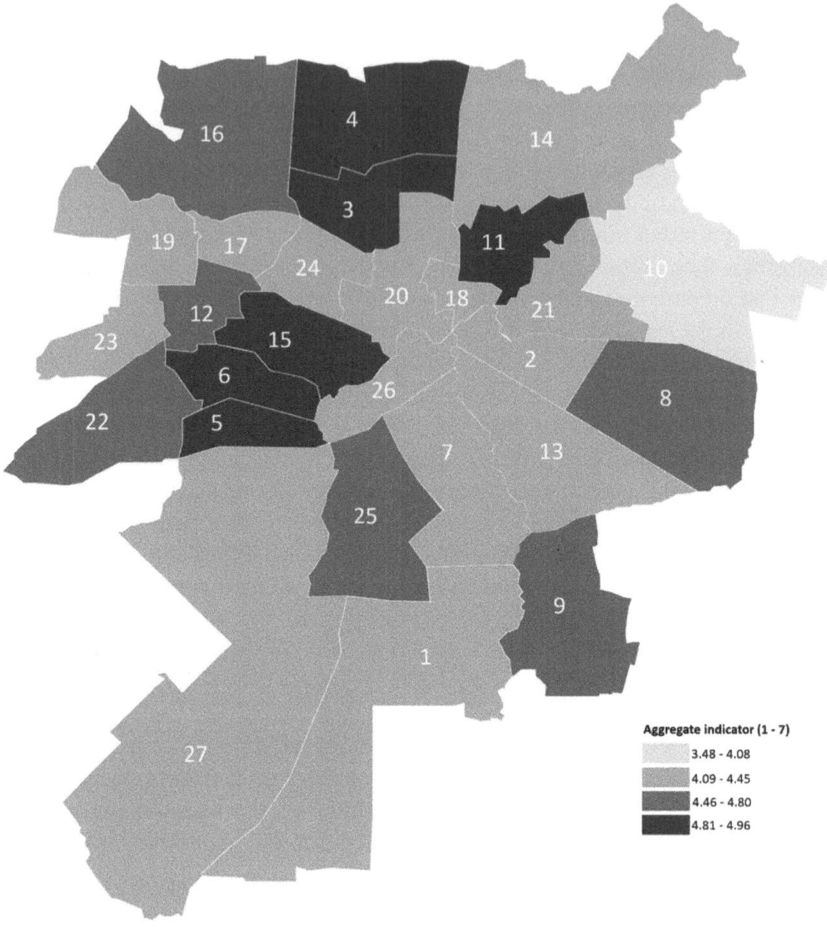

Notes: see table 1 for district names.

MAP 1 Aggregate indicators for neighborhood satisfaction in 27 districts in Lublin.

areas: the former are predominantly residential areas with housing estates accompanied by scattered, nondisturbing service buildings, while the latter constitute industrial areas supplemented by housing estates. The few southeastern districts that break out of this pattern include residential areas built within the last 20 years (districts no. 8 and 25), and a single-family suburban district incorporated as a result of the city's border expansion in the late 1980s (9).

The top-scoring districts were established during the communist era in the 1970s and 1980s, and feature moderately intensive housing development (usually 5-story buildings, less frequently 11-story buildings), complete and accessible amenities, and many green areas. Thanks to EU funding support for thermomodernization, prefabricated concrete housing estates have been renovated by adding insulation to their external walls and replacing old windows. Further, these areas have concentrated the majority of new residential investments in the last years. In contrast, two markedly backward districts (1 and 10) are scarcely populated peripheral areas with single-housing estates, relatively poor transport connections, and incomplete facilities. The urban center (districts 18 and 20), with townhouses erected largely in the 19th or early 20th century and rebuilt after World War II, performs somewhat below the urban average, featuring high scores for accessibility of public transport and condition of streets and pavements that are balanced by poor results for air quality, accessibility of parking space, and city cleanliness. On the whole, we find that the differences in neighborhood satisfaction are minor, which is partly related to the city's urban structure: Lublin can be described as a compact city with a single urban center and a multifunctional network of districts. This network is polycentric, yet distinct, with spatial proximity and good connections between districts. Hence, we argue that Lublin performs quite well in terms of making neighborhoods inclusive and resident-friendly, which is consistent with other assessments for Polish cities (OECD, 2017; Raszkowski & Bartniczak, 2019).

One important caveat to our interpretations is that subjective assessments might give a different picture of intra-urban differences than objective data, as the relationship between them is not very strong (Okulicz-Kozaryn, 2013; Okulicz-Kozaryn & Valente, 2019). Both approaches to the assessment of quality of life in urban areas are important and complementary (Costanza et al., 2007; Kothencz & Blaschke, 2017), yet the subjective approach has been receiving growing attention (Brezzi & Ramirez, 2016; OECD, 2013), particularly in the smart city literature (Macke, Casagrande, Sarate, & Silva, 2018). After all, such fundamental decisions by urban residents as their degree of citizen engagement or moving away from the city is driven more by what they perceive rather than what is really out there, and residents' perception of the quality of urban areas has been shown to be a relevant driver of happiness in general (Okulicz-Kozaryn & Valente, 2019). Nevertheless, objective data support the general conclusion—that intra-urban differences in Lublin are not large. Housing prices might be one good and accessible proxy for the overall quality of neighborhoods: the average price of a two-room apartment in the least attractive districts (districts 7, 10, 26, 27) is only twice as low as in the city center (20) and 40%–60% lower than in the top-scoring local areas (3, 4, 5, 6) (Lublin City Office, 2017).[b]

In contexts of pursuing smart governance and SDG 11, we highlight several arguments in favor of tools such as the one described in our paper. First, this is a very cost-effective way of conducting broad-based consultation, especially against traditional surveys or time-consuming consultation meetings. In the same vein, such online tools are characterized by high accessibility, which allows them to engage a large number of residents sharing their everyday

b. Another explanation for low intra-urban inequalities is that Lublin, like other Polish cities, has been ethnically homogeneous in the last half-century, thus avoiding ethnic segregation and the formation of ghettos that occurred in many cities in Western countries.

knowledge, thus increasing urban spatial intelligence. In addition, the approach addresses residents' perception of intra-urban inequalities—an issue that is difficult to explore empirically, yet remains a pressing urban problem affecting the urban structure and quality of life. Furthermore, information directly received from residents allows for more tailored, bottom-up investment planning in neighborhoods, and can assist local governments in prioritizing goals for local policies and determining the subsequent allocation of limited public resources. From the perspective of achieving SDGs, we argue that local authorities can use these tools to gather firm evidence on the kind of investments that are actually required to strengthen urban cohesion and to improve living conditions in somewhat disadvantaged districts. If carried out on a regular basis, such tools capture the dynamics of change in residents' perception of the quality of their space, and thus assist urban planners in their pursuit of more inclusive and sustainable urban areas. We also highlight the fact that the example discussed in this chapter is in synergy with other participation measures introduced by urban authorities. On the one hand, it is rooted in participatory online budget voting, which contributes to its effectiveness. On the other, it fosters urban collaboration and decentralization by providing robust data for district councils, which may be further used in local decision-making and local action plans for neighborhoods. On a more fundamental level, such tools may increase the transparency of decision-making and ensure trust in the public sector, thus contributing to a general culture of good governance.

5. Conclusions and future directions

The chapter contributed to our understanding of how urban authorities might implement "smart governance" by organizing the process of gathering new data from the public, and thus increase the rationality of the adopted decisions. It provides an example of how a more participatory and inclusive approach to recognize neighborhood satisfaction could contribute to the achievement of SDG 11. More specifically, it presented an application of a tool for recognizing residential satisfaction with urban public amenities in urban districts implemented in the city of Lublin (Poland), depicting it against the broader patterns of "smart city" development in Poland. Additionally, the chapter shed some new light on the debate on intra-urban inequalities within urban areas.

We gathered opinions expressed by almost 6000 participants (2% of the urban population) who took part in online voting on participatory budget projects. The results show that Lublin (presumably like other similar regional cities in Poland) is not affected by strong spatial inequalities in terms of public goods provision, while the existing, moderate differences concern mainly infrastructure for urban mobility, air quality, and sports facilities and amenities. Our aggregate indicator for neighborhood satisfaction ranged from 4.0 to 5.0 (on a 1–7 scale) for 25 out of 27 districts, while two markedly backward areas are sparsely populated, peripheral districts with scattered low-density single-housing estates.

We recognized some spatial patterns in neighborhood satisfaction and compared our results with insights from objective data. We also discussed how our tool might contribute to a culture of good governance as well as to sustainable development of urban areas emphasized in SDG 11.

Although our tool gathered new data from city dwellers, thus shedding some light on spatial differences in residential satisfaction, we recognize several issues with our approach and directions for the future. We considered issues with data quality, particularly surrounding its representativeness, and expressed skepticism about developing technologically advanced (though less inclusive) software tools. However, enhancing our spatial precision through a geo-survey could be a promising direction for the future. Linking questions with places on maps allows us both to conduct spatial analyses at a finer level of granularity and to calculate measures of accessibility of local amenities (see, e.g., Czepkiewicz et al., 2017; Haklay et al., 2018), whereas the expected decrease in residents' willingness to participate might not necessarily be substantial. Furthermore, we recommend that large-scale surveys, which only constitute a form of one-way communication, should be supplemented with tools that allow for two-way communication with fewer, albeit more advanced, users. In this respect, surveys cocreated by planners and the public that would allow for co-analysis of the resulting datasets could further ensure human-friendly urban planning. On a more fundamental level, the critical concern from the perspective of "smart governance" is how the government integrates the gathered data into its policies aimed at making cities more inclusive, safe, and sustainable, and how local authorities communicate the added value of this data to residents to further increase their civic engagement.

References

Ahvenniemi, H., Huovila, A., Pinto-Seppä, I., & Airaksinen, M. (2017). What are the differences between sustainable and smart cities? *Cities*, *60*, 234–245. https://doi.org/10.1016/j.cities.2016.09.009.

Albino, V., Berardi, U., & Dangelico, R. M. (2015). Smart cities: Definitions, dimensions, performance, and initiatives. *Journal of Urban Technology*, *22*(1), 3–21. https://doi.org/10.1080/10630732.2014.942092.

Anthopoulos, L., Janssen, M., & Weerakkody, V. (2019). A unified smart city model (USCM) for smart city conceptualization and benchmarking. In *Smart cities and smart spaces: Concepts, methodologies, tools, and applications* (pp. 247–264). IGI Global.

Barber, B. (2013). *If mayors ruled the world: Dysfunctional nations, rising cities*. New Haven, CT: Yale University Press.

Borsekova, K., & Nijkamp, P. (2018). Smart cities: A challenge to research and policy analysis. *Cities*, *78*, 1–3. https://doi.org/10.1016/j.cities.2018.03.015.

Brezzi, M., & Ramirez, M. D. (2016). Building subjective well-being indicators at the subnational level: A preliminary assessment in OECD regions. In *OECD regional development working papers, 2016/03*. https://doi.org/10.1787/5jm2hhcjftvh-en.

Caragliu, A., Del Bo, C., & Nijkamp, P. (2011). Smart cities in Europe. *Journal of Urban Technology*, *18*(2), 65–82. https://doi.org/10.1080/10630732.2011.601117.

Costanza, R., et al. (2007). Quality of life: An approach integrating opportunities, human needs, and subjective well-being. *Ecological Economics*, *61*(2–3), 267–276. https://doi.org/10.1016/j.ecolecon.2006.02.023.

Czepkiewicz, M., Jankowski, P., & Młodkowski, M. (2017). Geo-questionnaires in urban planning: Recruitment methods, participant engagement, and data quality. *Cartography and Geographic Information Science*, *44*(6), 551–567. https://doi.org/10.1080/15230406.2016.1230520.

Foray, D., David, P. A., & Hall, B. (2009). Smart specialisation—The concept. *Knowledge Economists Policy Brief*, *9*(85), 100.

Gawłowski, R., & Jarosz, A. J. (2019). Infrastruktura partycypacji społecznej. Pojęcie i wykorzystanie w polskich miastach [Infrastructure of public participation. The concept and its application in Polish cities]. *Studia Regionalne i Lokalne*, *4*(78), 75–94.

Giffinger, R., Fertner, C., Kramar, H., Kalasek, R., Pichler-Milanović, N., & Meijers, E. (2007). *Smart cities. Ranking of European medium-sized cities*. Vienna, UT: Centre of Regional Science. Retrieved from http://www.smart-cities.eu/download/smart_cities_final_report.pdf (May 2020).

Haklay, M., Jankowski, P., & Zwoliński, Z. (2018). Selected modern methods and tools for public participation in urban planning—A review. *Quaestiones Geographicae*, *37*(3), 127–149. https://doi.org/10.2478/quageo-2018-0030.

Knight Frank. (2018a). *Wroclaw—Towards a smart city*. Retrieved from https://content.knightfrank.com/research/1614/documents/en/wroclaw-towards-a-smart-city-2018-5863.pdf (May 2020).

Knight Frank. (2018b). *Warsaw—Towards a smart city*. Retrieved from https://content.knightfrank.com/research/1500/documents/en/warsaw-towards-smart-city-april-2018-5462.pdf (May 2020).

Kosovac, A., Acuto, M., & Jones, T. L. (2020). Acknowledging urbanization: A survey of the role of cities in UN frameworks. *Global Policy*, *11*(3), 293–304. https://doi.org/10.1111/1758-5899.12783.

Kothencz, G., & Blaschke, T. (2017). Urban parks: Visitors' perceptions versus spatial indicators. *Land Use Policy*, *64*, 233–244. https://doi.org/10.1016/j.landusepol.2017.02.012.

KPM. (2015). *Krajowa Polityka Miejska 2023* [National Urban Policy 2023]. Ministerstwo Infrastruktury i Rozwoju [Ministry of Infrastructure and Development]. Retrieved from https://www.funduszeeuropejskie.gov.pl/media/74967/Krajowa_Polityka_Miejska_2023.pdf (May 2020).

KSRR. (2019). *Krajowa Strategia Rozwoju Regionalnego 2030* [National Regional Development Strategy 2030]. Ministerstwo Infrastruktury i Rozwoju [Ministry of Infrastructure and Development]. Retrieved from https://www.gov.pl/web/inwestycje-rozwoj/krajowa-strategia-rozwoju-regionalnego (May 2020).

Lytras, M. D., Visvizi, A., Chopdar, P. K., Sarirete, A., & Alhalabi, W. (2021). Information Management in Smart Cities: Turning end users' views into multi-item scale development, validation, and policy-making recommendations. *International Journal of Information Management*, *56*, 102146. https://doi.org/10.1016/j.ijinfomgt.2020.102146.

Lytras, M. D., Visvizi, A., & Sarirete, A. (2019). Clustering smart city services: Perceptions, expectations, responses. *Sustainability*, *11*(6), 1669. https://doi.org/10.3390/su11061669.

Macke, J., Casagrande, R. M., Sarate, J. A. R., & Silva, K. A. (2018). Smart city and quality of life: Citizens' perception in a Brazilian case study. *Journal of Cleaner Production*, *182*, 717–726. https://doi.org/10.1016/j.jclepro.2018.02.078.

Maleszyk, P. (2019). *Czego potrzebuje Lublin? Raport z badania jakości życia w dzielnicach Lublina* [What Lublin needs? Report on quality of life in Lublin districts]. Lublin: Urząd Miasta Lublin [Lublin City Office]. Retrieved from https://depot.ceon.pl/handle/123456789/16547 (May 2020).

Maleszyk, P., & Sagan, M. (2018). *Lublin 2030–europejska metropolia?* [Lublin 2030 – European metropolis?]. Lublin: Urząd Miasta Lublin [Lublin City Office].

Manville, C., Cochrane, G., Cave, J., Millard, J., Pederson, J. K., Thaarup, R. K., … Kotterink, B. (2014). *Mapping smart cities in the EU*. European Parliament.

Marcińczak, S. (2012). The evolution of spatial patterns of residential segregation in Central European Cities: The Łódź Functional Urban Region from mature socialism to mature post-socialism. *Cities*, *29*(5), 300–309. https://doi.org/10.1016/j.cities.2011.08.008.

Meijer, A., & Bolívar, M. P. R. (2016). Governing the smart city: A review of the literature on smart urban governance. *International Review of Administrative Sciences*, *82*(2), 392–408. https://doi.org/10.1177/0020852314564308.

Ministry of Digitalisation. (2020). *Innovation map*. Retrieved from https://www.gov.pl/web/cyfryzacja/mapa-innowacji (May 2020).

Mora, L., Bolici, R., & Deakin, M. (2017). The first two decades of smart-city research: A bibliometric analysis. *Journal of Urban Technology*, *24*(1), 3–27. https://doi.org/10.1080/10630732.2017.1285123.

Nam, T. (2012). Suggesting frameworks of citizen-sourcing via Government 2.0. *Government Information Quarterly*, *29*(1), 12–20. https://doi.org/10.1016/j.giq.2011.07.005.

Neirotti, P., De Marco, A., Cagliano, A. C., Mangano, G., & Scorrano, F. (2014). Current trends in smart city initiatives: Some stylised facts. *Cities*, *38*, 25–36. https://doi.org/10.1016/j.cities.2013.12.010.

OECD. (2013). *OECD guidelines on measuring subjective well-being*. OECD Publishing. https://doi.org/10.1787/9789264191655-en.

OECD. (2017). *The 2030 sustainable development agenda: Towards a successful implementation by Poland*. OECD Publishing. https://doi.org/10.1787/9789264287341-en.

Okulicz-Kozaryn, A. (2013). City life: Rankings (livability) versus perceptions (satisfaction). *Social Indicators Research*, *110*(2), 433–451. https://doi.org/10.1007/s11205-011-9939-x.

Okulicz-Kozaryn, A., & Valente, R. R. (2019). Livability and subjective well-being across European cities. *Applied Research in Quality of Life*, *14*(1), 197–220. https://doi.org/10.1007/s11482-017-9587-7.

Pistelok, P., & Martela, B. (2019). *Partycypacja publiczna. Raport o stanie polskich miast* [Report on state of Polish cities]. Warszawa/Kraków: Instytut Rozwoju Miast i Regionów.

Polanska, D. V. (2018). Going against institutionalization: New forms of urban activism in Poland. *Journal of Urban Affairs*, 1–12. https://doi.org/10.1080/07352166.2017.1422982.

Raszkowski, A., & Bartniczak, B. (2019). On the road to sustainability: Implementation of the 2030 agenda sustainable development goals (SDG) in Poland. *Sustainability*, *11*(2), 366. https://doi.org/10.3390/su11020366.

Ruhlandt, R. W. S. (2018). The governance of smart cities: A systematic literature review. *Cities*, *81*, 1–23. https://doi.org/10.1016/j.cities.2018.02.014.

Sikora-Fernandez, 2017Sikora-Fernandez, D. (2017). Strategia rozwoju miasta inteligentnego–perspektywa polska [Smart City development strategy—The polish perspective]. *Przedsiębiorczość i Zarządzanie*, *18*(2.3), 13–25.

Sikora-Fernandez, D. (2018). Smarter cities in post-socialist country: Example of Poland. *Cities*, *78*, 52–59. https://doi.org/10.1016/j.cities.2018.03.011.

SOR. (2017). *Strategia na rzecz Odpowiedzialnego Rozwoju* [Strategy for Responsible Development]. Warszawa [Warsaw]: Ministerstwo Rozwoju [Ministry of Development]. Retrieved from https://www.gov.pl/documents/33377/436740/SOR.pdf (May 2020).

Tomor, Z., Meijer, A., Michels, A., & Geertman, S. (2019). Smart governance for sustainable cities: Findings from a systematic literature review. *Journal of Urban Technology*, *26*(4), 3–27. https://doi.org/10.1080/10630732.2019.1651178.

UML. (2017). *Rynek nieruchomości w Lublinie. Informacja dla inwestorów* [Housing market in Lublin. Information for investors]. Retrieved from http://www.outsourcingportal.eu/pl/userfiles/file/_stale/pliki_firmy/rynek-nieruchomosci-w-lublinie-informacja-dla-inwestorow.pdf (May 2020).

United Nations. (2018). *2018 revision of world urbanization prospects*. Retrieved from https://www.un.org/development/desa/publications/2018-revision-of-world-urbanization-prospects.html (May 2020).

Visvizi, A., & Lytras, M. (2018). Rescaling and refocusing smart cities research: From mega cities to smart villages. *Journal of Science and Technology Policy Management*. https://doi.org/10.1108/JSTPM-02-2018-0020.

Visvizi, A., & Lytras, M. D. (2019a). Smart cities research and debate: What is in there? In *Smart cities: Issues and challenges* (pp. 1–14). Elsevier. https://doi.org/10.1016/B978-0-12-816639-0.00001-6.

Visvizi, A., & Lytras, M. D. (2019b). Sustainable smart cities and smart villages research: Rethinking security, safety, well-being, and happiness. *Sustainability*, *12*(1), 1–4. https://doi.org/10.3390/su12010215.

Visvizi, A., Matysek-Jedrych, A., & Mroczek-Dąbrowska, K. (2020). *Poland in the single market: Politics, economics, the euro*. London: Routledge. https://doi.org/10.4324/9780429285301.

Chapter 12

Cybersecurity, sustainability, and resilience capabilities of a smart city

Roberto O. Andrade[a], Sang Guun Yoo[a,b], Luis Tello-Oquendo[c], and Iván Ortiz-Garcés[d]

[a]*School of Systems Engineering, National Polytechnic School, Quito, Ecuador,* [b]*Smart Lab, National Polytechnic School, Quito, Ecuador,* [c]*College of Engineering, Universidad Nacional de Chimborazo, Riobamba, Ecuador,* [d]*Faculty of Engineering and Applied Sciences, University of the Americas, Quito, Ecuador*

1. Cybersecurity in context of smart cities and the SDGs

In 2015, the United Nations Development Program (UNDP) with its member states proposed a set of 17 Global Goals or Sustainable Development Goals (SDGs) in order to seek the eradication of poverty, the protection of the planet, gender equality, and the well-being of people. The SDG number 11 mentions that states and cities must be safe, resilient, sustainable, and inclusive (UNDP, 2015). Cities need to invest in housing, transport, and urban management; besides, they must establish strategies related to emissions management, water usage, human rights, and gender diversity. Achieving this sustainability involves some challenges for cities due to the high migration of people from rural to urban areas and continued population growth. Some interesting facts related to urban areas are the following:

- The people living in urban areas increased from 4.5 billion in 2015 to 6.5 billion in 2020.
- About 80% of the gross domestic product (GDP) is generated in urban areas.
- About 60%–80% of the planet's electrical energy is consumed in urban areas.
- The number of cities with more than 10 million inhabitants is estimated to grow from 10 cities in 1990 to 41 cities in 2030.

Several cities have chosen using a management model focused mainly on improving urban planning processes to continuously evaluate the resources available so that all citizens have equitable access and an adequate level of quality of life. The adoption of data-driven planning and decision-making

Smart Cities and the UN SDGs. https://doi.org/10.1016/B978-0-323-85151-0.00012-9

has taken an increasingly important role in cities, and the use of emerging technologies such as Internet of Things (IoT), BigData, and machine learning is making this adoption faster.

The management model, based on the use of data, technology, and reuse of resources in cities, has promoted the concept of *smart city*. Nowadays, several cities such as Paris, Amsterdam, and Tokyo have been classified as smart cities. This fact has motivated different international organizations such as the International Telecommunication Union (ITU), the International Organization for Standardization (ISO), and the Institute of Electrical and Electronics Engineers (IEEE) to begin working on recommendations for the implementation, monitoring, and continuous improvement of the aspects that make up the smart city (Andrade & Yoo, 2019). Maintaining the sustainability and resilience capabilities requires that the smart city adapts to changes generated by social, political, environmental, and natural factors. The smart city also should provide security to both its citizens and the critical infrastructures that support the provision of critical services such as energy, water, communications, education, health, among others (ITU, 2015).

Emerging technologies allow a smart city to manage the aspects of sustainability and resilience (Visvizi & Lytras, 2019a). Information from multiple sensors installed in different smart city components is used; then, large volumes of data generated for these sensors are analyzed. The smart city has been exposed to cyberattacks, as indicated in the report of the World Economic Forum (WEF) related to the top 10 risks worldwide. Based on the probability of occurrence, the 2017 WEF report placed cyberattacks in the sixth position of the list. In the 2019 report, it was located in the fifth position (WEF, 2020); therefore, the criticality and probability of occurrence of cybersecurity attacks have increased worldwide.

This chapter examines the connection between cybersecurity and smart cities' sustainability and resilience capacity. The argument is structured as follows. Section 2 maps the key cybersecurity threats and their impact on smart city domains such as transportation, health, home, and others. Then, Section 3 analyzes the cybersecurity aspects and how cyberattacks impact smart city domains; it is evaluated considering each layer of the smart sustainability city architecture proposed by the International Telecommunication Union (ITU). Section 4 studies how these cyberattacks impact the sustainability and resilience capabilities of the cities. Finally, Section 5 discusses how the city could exploit cybersecurity aspects to protect its sustainability and resilience capabilities from cyberattacks.

2. Cybersecurity aspects in the smart city

The ITU (2015) establishes the architecture for smart and sustainable cities, as illustrated in Fig. 1. Cyberattacks could be present in each layer of this architecture.

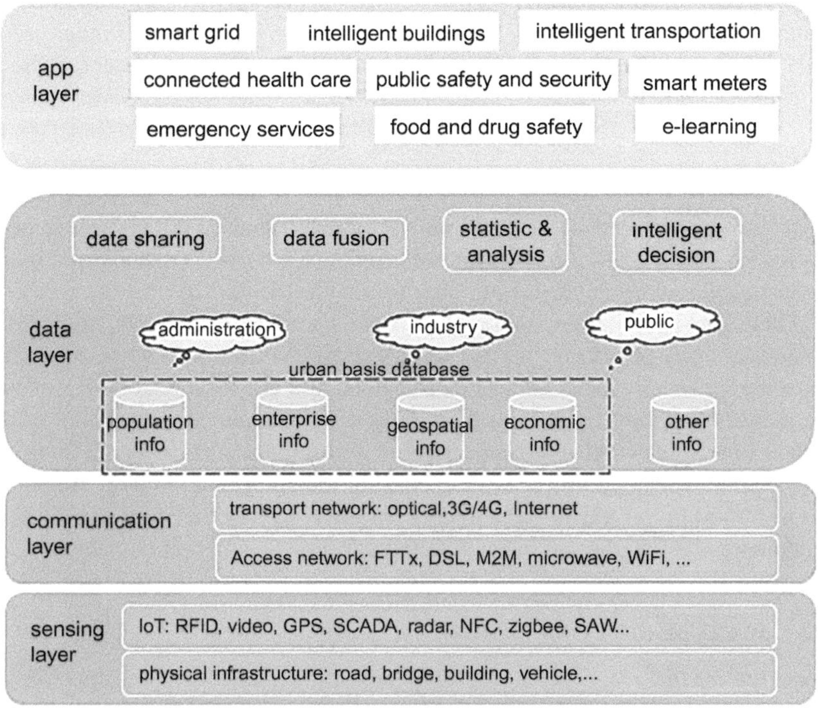

FIG. 1 Proposed architecture for smart and sustainable cities.

The failure of infrastructure in a smart city can put citizens at risk, and the increment of cyberattacks to cities in recent years elevates this risk even more (Schaffers et al., 2011). For instance, the city of Dublin has ransomware attacks on over 3000 user's records, while the city of Stockholm suffered jamming attacks in air traffic control, and the city of Johannesburg had a shakedown to power plants (BBC News, 2020; Breaking News Ireland, 2019). The different infrastructures and services in the different domains of the smart city may present vulnerabilities and expose the smart city to cyberattacks. In what follows, we describe several smart city applications and how the attackers could exploit vulnerabilities to take advantage of the situation.

2.1 Smart grid cybersecurity

A smart grid contributes to the objective of increasing efficiency, reducing costs, and improving the environmental impact produced by energy consumption that is between 60% and 80% in cities. By including smart sensors and meters, it is possible to generate a real-time analysis of energy consumption. Attackers can exploit SCADA systems' vulnerabilities to alter the exchange of data between energy plants and end-users, aiming at causing an interruption in the energy service delivery. At the end-user device level, attackers may attempt to perform fraudulent processes to modify the consumption or power values of other users. SANS (2020) mentions that some attacks on SCADA systems include: malicious firmware development, SCADA Hijack, Kill-Disk overwrite, and power outage.

2.2 Smart transportation cybersecurity

Smart transportation enables solutions to improve urban mobility, reduce congestion on public transport, control pollution, and improve citizens' quality of life, especially in urban areas with high vehicle traffic density.

Some threats to smart transportation are unauthorized access, data theft, vandalism, data alteration, among others (ENISA, 2020). Disruption of the transport service by attackers can cause social and economic damage; citizens use public transport to carry out their daily activities, and if they cannot travel to their jobs, this can generate an economic impact. An attacker could interrupt the electrical flow and power source, causing transport units to stop. Attackers can also modify the traffic signs located on the streets or transport terminals to generate confusion, disrupt ticket vending machines, and other damages.

2.3 Smart health cybersecurity

Telemedicine has a significant impact on providing health support, especially in the case of natural disasters or where remote access is required. A concern in the hospital domain is that patient privacy may be exposed if an attacker accesses medical records.

One of the most common attacks is ransomware, which has the following pattern: attackers gain access to information systems through direct use of USB, phishing email, theft of medical devices, or software vulnerabilities. Then, they execute malicious code that encrypts the "time sensitivity" feature information to request a payment. Another common type of attack is information theft to get critical information from citizens and sell it on the black market (Harvard, 2020).

2.4 Public safety cybersecurity

Surveillance systems are installed in the city to have real-time capabilities to identify street violence or theft. Attackers can access surveillance systems and control the visualization of the city spaces or block the cameras to commit vandalism or crimes. On the other hand, there is a concern about aspects of citizen privacy when having systems throughout the city and that these cannot be used unethically. Sometimes users are not familiar with security issues, and they become perfect targets for attackers when they interact with many smart city services through their smartphones, tablets, and computers, revealing personal data such as gender, age, and location (Vattapparamban, Güvenç, Yurekli, Akkaya, & Uluağaç, 2016).

Drones play a vital role in increasing public safety capacity by exploring areas that are difficult to reach and providing an area of rapid coverage and connectivity.

However, the pervasive use of drones for future smart cities also brings together several technical and social concerns and challenges that need to be addressed, including cybersecurity, privacy, and public safety. The communication between the drone and ground station controller is typically performed by the WiFi network, which is vulnerable to security breaches. Another type of communication in drones includes GPS communication; consequently, GPS spoofing attack is commonly performed on drones (Zhang, Li, Du, & Dou, 2020).

Intelligent traffic lights attack to cause potential traffic disruption and public safety degradation, cybersecurity posture in traffic light systems allows the traffic management authority to determine and prioritize measures to guard against various cyber threats, to ensure safety and order in traffic infrastructure, security concerns must be addressed when designing traffic systems.

2.5 Wireless communications and hotspots cybersecurity

Smart cities offer free wireless hotspots for public WiFi. Without encryption, the plain text data pass unprotected through the air, where it can be intercepted by cybercriminals that would try to capture valuable personal information. WiFi security threats are known as man-in-the-middle attacks. Attackers will intercept messages between two users who believe they are talking directly to one another. Evil twins are designed to look and act exactly like a legitimate access point (AP). Hackers can clone a known and trusted AP and create an identical one. The most common soft access points are hotspots. Users sometimes forget to turn these hotspots off, so the next time they connect to a public WiFi network, the hotspot creates a vulnerable entry point that hackers can exploit. These hotspots can create a serious issue, even if the network is otherwise secure.

2.6 Smart home cybersecurity

The smart home allows its occupants to have information and control of household components such as light levels, air quality, and electronic equipment. The entire set of sensors and connected components generates information about the habits of household members. The interconnection of devices at home, raises privacy concerns since an attacker could access sensitive information, or even the operators themselves could sell the data to third parties with consumer preferences and habits gathered from users.

A report by the Federal Bureau of Investigation (FBI) in November 2019 mentions that attackers can use the smart TV as a gateway to access the resources at home to steal information or invade the privacy of their occupants (FBI, 2020).

3. Attacks on the components of the smart city

The smart city solutions, as described above, are made up of different technological components specified in the smart, sustainable city architecture proposed by the ITU and previously described in Fig. 1. Cyberattacks can exploit the vulnerabilities that these components can present and affect confidentiality, integrity, and availability in the smart city. Nautiyal, Malik, and Agarwal (2018) mention that some security challenges faced by smart cities are due to aspects such as:

- data leakage of personal information,
- insecure hardware due to lack of standardization and security testing protocols,
- larger attack surface, due to a large number of devices, and
- bandwidth consumption by a large number of sensors.

In the following, we analyze the cybersecurity aspects present in each layer of the architecture proposed by the ITU. This will allow providing an overview of the security situation that will be useful when establishing defense strategies.

3.1 Attacks on application layer

The attacker will try to obtain access through multiple techniques, such as SQL injection or cross-site scripting (XSS), using the application layer. Distributed Denial of Service (DDoS) is another typical attack in the application layer; the attacker uses legitimate requests to exhaust application resources (Praseed & Thilagam, 2019). Single sign-on (SSO) are great ways to share technology across platforms, even is possible to use credentials for email or social networks, but we must ensure that the credentials cannot be used to abuse permissions not expressly granted to a third-party service. Attacks to SSO protocols can abuse a lack of a single security check and cross-phase (Mainka, Mladenov, Schwenk, & Wich, 2017). Therefore, it is necessary to request the providers that their service allows specific permissions to be configured. In some cases, revoking the service to a specific user creates a short window of exposure of the data that allows continuing access later that the authorization period ends.

3.2 Attacks on the data layer

Security on the cloud is a shared responsibility for providers and users. The cloud provider should implement the appropriate measures to reduce the probability that an attacker and users are responsible for security configuration. Cloud solutions, especially commercial ones, have a set of security mechanisms such as firewall, multiple-factor authentication, and role-based access control (RBAC), but their effectiveness depends on a correct configuration that depends mostly on the user's expertise. An adversary can compromise the resources of a cloud service provider; they could access the central system and obtain a privileged position to inspect and modify running virtual machine (VM) systems.

In the Infrastructure as a Service (IAAS) computing infrastructure, the isolation between different customers' co-residing VMs is a critical security issue. It is possible to infer secret keys from another guest VM through a side-channel attack (Rakotondravony, Taubmann, & Mandarawi, 2017), consuming excessive resources on the physical server to force a target VM to migrate to a physical server over which the attacker has extra control. Once access is acquired, the adversary must maintain it long enough to identify which system contains the cloud resource relevant to their interests. It is possible to collect cloud metrics that reveal anomalies in the behavior of particular cloud resources, but sometimes is difficult to detect an intrusion since an attacker with sufficient privileges; in some cases, the attack can occur during the migration processes of the VMs by internal attackers, this difficult the attack detection process (Duncan, Creese, Goldsmith, & Quinton, 2013).

3.3 Attacks on the communication layer

Attacks to the infrastructure of the Internet associated with Internet protocols, such as the hijacking of the Border Gateway Protocol (BGP), attackers can directly originate the bogus routes of the prefixes, or even stealthier, further spoof the autonomous system (AS) paths of the routes to make them appear to be originated by others (Qiu, Gao, Ranjan, & Nucci, 2007).

Attackers are looking for a poorly designed communication layer, which allows bypassing authentication mechanisms and injecting commands into interface scripts.

3.4 Attacks on the sensing layer

In the sensing layer, the IoT network is the most used technology to capture information in the smart city. The botnet, man-in-the-middle attacks, social engineering, data and identity defeats, and Denial of Service (DoS) attacks are the most common attacks in this layer (Okul & Ali, 2017).

Attackers try to use man-in-the-middle attacks to compromise the peripheral device because they are the most exposed to find vulnerabilities to allow an attacker to capture the login credentials or tokens in the communication model. One of the most complex aspects of managing peripheral devices is that they are distributed in all IoT networks. Thus, it is necessary to monitor the behavior of both the devices, the communication, and the application layer (Eigner, Kreimel, & Tavolato, 2016).

4. Cybersecurity, sustainability, and resilience

Cities face a growing scale of adversity and challenges in the 21st century, from the effects of climate change up to the growth of migrant populations, inadequate infrastructure, pandemics, and cyberattacks. Building resilience and sustainability spaces aim to take a comprehensive look at a city: to understand the systems that build the cities and the risks they may face.

Nautiyal et al. (2018) mention that not only the technical aspects jointly with the economic, environmental, and social aspects of the organization must be considered in cybersecurity administration. From the perspective of the smart city, two additional relevant aspects are the city's sustainability and resilience.

Sustainability can be established as the city's capacity to make efficient use of its resources for the current and future population. Microsoft Corporation (2020) mentions that they will be essential for the digital transformation of cities and the advancement of urban cyber resilience: cloud computing, the IoT, big data, machine learning, and artificial intelligence. The path to cyber resilience is not an easy one. Understanding cyber resilience and developing strategies to address it requires cities to think, organize, and operate differently.

According to ISO 22316:2017, "Security and resilience," the resilience is defined as the city's ability to adapt to adverse events that affect its regular operation. The resilience of the smart city is a relevant aspect to maintain its ability to be sustainable; resilience should be aligned with the objectives of sustainability. To adapt the organization concept of the smart city proposed by ISO 22316, it should consider strengthening resilience as a strategic objective, for which ISO 22316 recommends covering the following aspects:

- ability to anticipate and address risks and vulnerabilities;
- coordination and integration of management disciplines to improve coherence and performance; and
- understanding of interested parties and dependencies that support strategic goals and objectives.

Fig. 2 illustrates that any aspect that affects smart city activities' regular operation can directly or indirectly affect its sustainability and resilience capacities. The attack on infrastructures that allow the supply of water or electricity could generate a social, economic, and environmental impact.

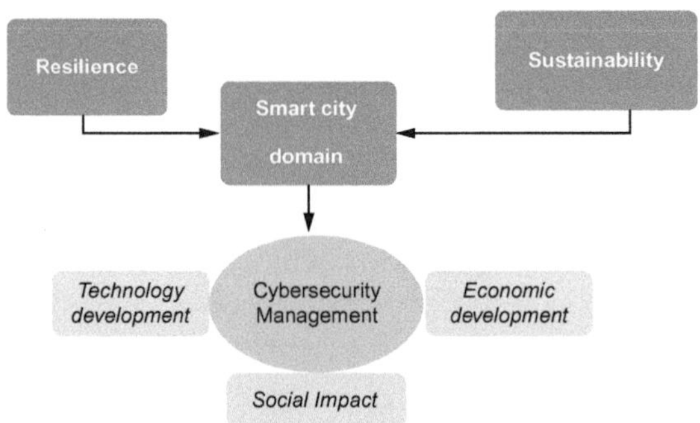

FIG. 2 The relation between cybersecurity management and the resilience and sustainability of a smart city.

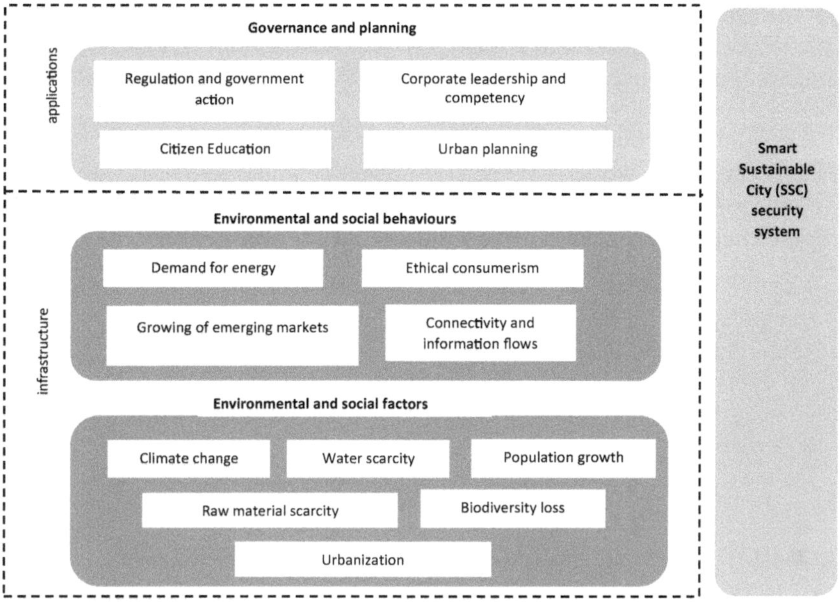

FIG. 3 Challenges of smart cities to maintain sustainability.

Silva, Khan, and Han (2018) present some trends that are represented in Fig. 3, related to the challenges that smart cities face in maintaining sustainability.

The most relevant cybersecurity aspects in smart cities that we have identified based on the conducted literature review are:

- privacy;
- cyber-physical systems security;
- security in critical infrastructures; and
- security applied to the different domains of the city, smart home, smart health, smart transportation, and smart governance.

As cities plan for resilience, they realize that it depends on the proper functioning of information and communication technologies (ICTs) in an adaptive technological infrastructure. This cyber resilience is demonstrated by the preparedness of a city's ICT infrastructure for a crisis, its responsiveness, and its ability to reinvent its ICT structure in sustained stress and acute disruptions. It should be considered essential aspects, such as creating clear goals and objectives, identifying people and services that need to be included in the planning when building cyber resilience in a city.

These goals and objectives must be practical to avoid cyber-crime damage. Note that it will cost the world $6 billion annually by 2021; it is exponentially more than the damage inflicted by natural disasters in one year and more profitable than the global trade of all the main illegal drugs combined; only New York City could face more than $2.3 billion in losses to its GDP due to cyber-related attacks in 2020. Considering the mentioned facts, we can realize that resilience is a crucial part of the growth of a smart city, providing resources, experience, and a framework to "help cities around the world be more resistant" (Ch, Gadekallu, Abidi, & Al-Ahmari, 2020).

Cities must have a digital security mindset to protect themselves, says the WEF, because "the world is in the early stages of a technological war against city governments and urban infrastructure." With this in mind, Microsoft (2017) proposes five steps to develop an e-resilience strategy as follows:

1. Identify key threats and assess their impact on critical cyber systems and functions;
2. Classify and prioritize critical services;
3. Set cyber resilience goals and objectives;
4. Develop the desired cyber resistance results and identify and test capabilities to achieve them; and
5. Define roles, responsibilities, and determine the necessary resources.

5. On the management of cybersecurity in smart city

Cyberattacks can affect the sustainability and resilience aspects of cities. Depending on the attack, it could generate social, economic, and environmental implications. How each scenario is handled depends on the IoT service provider and how the detected incidents are tackled. Fig. 4 indicates the factors to develop cybersecurity guidelines in a smart city context. Based on these factors, we propose the following recommendations to build effective cybersecurity management:

- define a security infrastructure for systems exposed to the public Internet;
- define an incident response model;

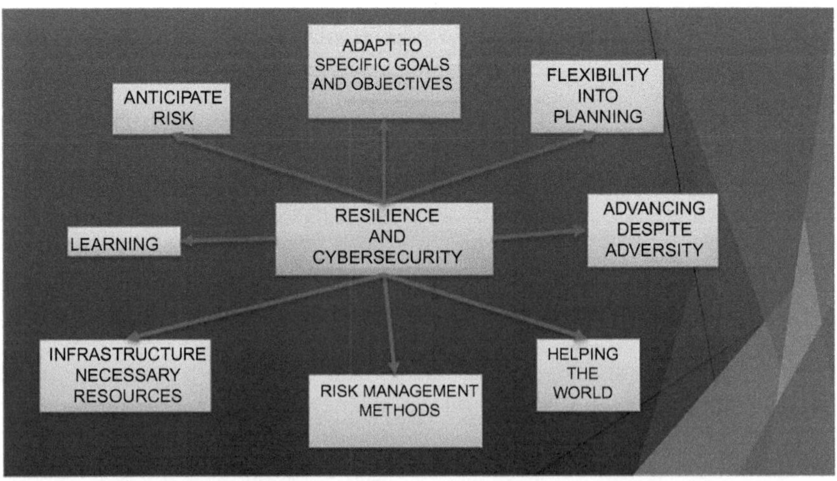

FIG. 4 Factors to develop cybersecurity guidelines in a smart city context.

- define a recovery model;
- increase authentication through the services ecosystem; and
- define authentication and authorization models.

One of the relevant cybersecurity management activities is to assess the risks to which the different components of the smart city are exposed. Cybersecurity risk analysis should consider factors such as:

- the likelihood of events and consequences;
- the nature and magnitude of consequences;
- the complexity and connectivity;
- the time-related factors and volatility;
- the effectiveness of existing controls; and
- the sensitivity and confidence levels.

The risk analysis may be influenced by the divergence of opinions, biases, perceptions of risk, and judgments. Additional influences are the quality of the information used, the assumptions and exclusions made, any limitations of the techniques, and how they are executed. These influences should be considered, documented, and communicated to decision-makers. Highly uncertain events can be challenging to quantify. This can be an issue when analyzing events with severe consequences. In such cases, using a combination of techniques generally provides greater insight.

Risk analysis provides input to (i) risk evaluation, (ii) decisions on whether risk needs to be treated and how, and (iii) find the most appropriate risk treatment strategy and methods. The results provide insight for decisions, where choices are being made, and the options involve different types and levels of risk.

6. Discussion

We have repeatedly emphasized the importance of the smart city to make the world a more sustainable place. Cities that reduce the negative environmental impact, correctly analyze and manage the large volumes of data. According to some experts in urban regeneration, cities of the future are more resilient, which will progressively replace the concept of sustainability in the coming years (Visvizi & Lytras, 2019b). With the advancement of technology, useful applications are developed, but attacks that harm a person's stability or society are also developed.

Cities could be prepared to respond to different types of threats (wars, natural disasters, economic crises, depletion of natural resources, among others events), and also could be capable of quickly recovering from the impacts suffered by the system, causing the least possible damage to public health, economy, and security. When these unexpected events occur, service interruptions and reductions result in the need for personnel reallocation, operation mode, and financial resources. Resilience has emerged as an essential framework to better prevent and respond to service disruptions within which threat planning should be used.

Cyberattacks on transportation or smart grid systems could produce that city to stop its operation, and even more, that city could not return to normalcy in a short time. This event could associate economic losses and social impact. Therefore, resilience is the best way to progress since a society aims to adapt and grow to have a better economy and thus provide a better quality of life for its inhabitants, without neglecting cyberattacks and never underestimating attackers. Today, the advent of the Industrial Internet of Things poses new security challenges. However, industrial control systems, for example, are designed to be resilient, capable of recovering from process failures with limited impact on operations, security challenges advocate for a new risk assessment, and security strategies. The inclusion of cognitive sciences (fuzzy engineering, Bayesian networks, convolutional networks) in cybersecurity could be considered to evaluate and predicted attacks, threats, and risks more adaptively and dynamically.

References

Andrade, R. O., & Yoo, S. G. (2019). A comprehensive study of the use of LoRa in the development of smart cities. *Applied Sciences*, *9*, 4753.

BBC News. (2020). *Ransomware hits Johannesburg electricity supply*. Available from https://www.bbc.com/news/technology-49125853. (Accessed 15 July 2020).

Breaking News Ireland. (2019). *Over 3000 Luas users may have had records compromised in cyberattack*. Available from https://www.breakingnews.ie/ireland/over-3000-luas-users-may-have-had-records-compromised-in-cyber-attack-895306.html. (Accessed 15 July 2020).

Ch, R., Gadekallu, T. R., Abidi, M. H., & Al-Ahmari, A. (2020). Computational system to classify cyber crime offenses using machine learning. *Sustainability*, *12*, 4087.

Duncan, A., Creese, S., Goldsmith, M., & Quinton, J. S. (2013). Cloud computing: Insider attacks on virtual machines during migration. In *12th IEEE international conference on trust, security, and privacy in computing and communications, Melbourne, VIC, 2013* (pp. 493–500).

Eigner, O., Kreimel, P., & Tavolato, P. (2016). Detection of man-in-the-middle attacks on industrial control networks. In *2016 international conference on software security and assurance (ICSSA), St. Polten, 2016* (pp. 64–69).

ENISA. (2020). *Smart transport*. Available from https://www.enisa.europa.eu/topics/iot-and-smart-infrastructures/smart-transport?tab=publications. (Accessed 15 July 2020).

FBI. (2020). *Oregon FBI tech tuesday: Securing smart TVs*. Available from https://www.fbi.gov/contact-us/field-offices/portland/news/press-releases/tech-tuesdaysmart-tvs. (Accessed 16 July 2020).

Harvard. (2020). *State of cybersecurity & cyber threats in healthcare organizations*. Available from https://blogs.harvard.edu/cybersecurity/files/2017/01/risks-and-threats-healthcare-strategic-report.pdf. (Accessed 15 July 2020).

ITU. (2015). *Technical report on: "Cybersecurity, data protection and cyber resilience in smart sustainable cities"*. Available from https://www.itu.int/md/T13-SG05-151012-TD-GEN-1005/en. (Accessed 15 July 2020).

Mainka, C., Mladenov, V., Schwenk, J., & Wich, T. (2017). SoK: Single sign-on security—An evaluation of OpenID connect. In *2017 IEEE European symposium on security and privacy (EuroS&P), Paris, 2017* (pp. 251–266).

Microsoft. (2017). *Cyber resilience: digitally empowering cities [online]*. Available at: https://query.prod.cms.rt.microsoft.com/cms/api/am/binary/RW6auc. (Accessed 31 December 2020).

Microsoft Corporation. (2020). *Cyber resilience: Digitally empowering cities*. Available from https://query.prod.cms.rt.microsoft.com/cms/api/am/binary/RW6auc. (Accessed 15 July 2020).

Nautiyal, L., Malik, P., & Agarwal, A. (2018). Cybersecurity system: An essential pillar of smart cities. In *Computer communications and networks* (pp. 25–50). Springer. https://doi.org/10.1007/978-3-319-76669-0_2.

Okul, Ş., & Ali, A. M. (2017). Security attacks on IoT. In *International conference on computer science and engineering (UBMK), Antalya, 2017* (pp. 1–5).

Praseed, A., & Thilagam, P. S. (2019). DDoS attacks at the application layer: Challenges and research perspectives for safeguarding web applications. *IEEE Communication Surveys and Tutorials*, *21*(1), 661–685 (First quarter, 2019).

Qiu, J., Gao, L., Ranjan, S., & Nucci, A. (2007). Detecting bogus BGP route information: Going beyond prefix hijacking. In *Third international conference on security and privacy in communications networks and the workshops—SecureComm 2007, Nice, France, 2007* (pp. 381–390).

Rakotondravony, N., Taubmann, B., & Mandarawi, W. (2017). Classifying malware attacks in IaaS cloud environments. *Journal of Cloud Computing*, *6*, 26. https://doi.org/10.1186/s13677-017-0098-8.

SANS. (2020). *Analysis of the cyber attack on the Ukrainian power grid*. Available from https://ics.sans.org/media/E-ISAC_SANS_Ukraine_DUC_5.pdf. (Accessed 15 July 2020).

Schaffers, H., Komninos, N., Pallot, M., Trousse, B., Nilsson, M., & Oliveira, A. (2011). Smart cities and the future internet: Towards cooperation frameworks for open innovation. In J. Domingue, et al. (Eds.), *Lecture notes in computer science: Vol. 6656. The future internet. FIA 2011*. Berlin, Heidelberg: Springer.

Silva, B. N., Khan, M., & Han, K. (2018). Towards sustainable smart cities: A review of trends, architectures, components, and open challenges in smart cities. *Sustainable Cities and Society*, *38*, 697–713. https://doi.org/10.1016/j.scs.2018.01.053.

UNDP. (2015). *Goal 11: Sustainable cities and communities*. Available from https://www.undp.org/content/undp/en/home/sustainable-development-goals/goal-11-sustainable-cities-and-communities.html.

Vattapparamban, E., Güvenç, İ., Yurekli, A.İ., Akkaya, K., & Uluağaç, S. (2016). Drones para ciudades inteligentes: Problemas de ciberseguridad, privacidad y seguridad pública. In *Conferencia internacional de comunicaciones inalámbricas e informática móvil (IWCMC)* (pp. 216–221). IEEE.

Visvizi, A., & Lytras, M. (Eds.). (2019a). *Smart cities: Issues and challenges: Mapping political, social and economic risks and threats* Elsevier.

Visvizi, A., & Lytras, M. D. (Eds.). (2019b). *Politics and technology in the post-truth era*. Bingley: Emerald Publishing.

World Economic Forum. (2020). *The global risks report 2020*. Available from https://www.weforum.org/reports/the-global-risks-report-2020. (Accessed 1 May 2020).

Zhang, H., Li, P., Du, Z., & Dou, W. (2020). Risk entropy modeling of surveillance camera for public security application. *IEEE Access*, *8*, 45343–45355. https://doi.org/10.1109/ACCESS.2020.2978247.

Part IV

Sociological reflection on the smart city and beyond

Chapter 13

Sociological reflections on smart cities and city times

Pablo de Gracia-Soriano[a,b], Diana Jareño-Ruiz[b], María Jiménez-Delgado[b], and Antonio Jiménez-Delgado[b]
[a]*Complutense University of Madrid, Madrid, Spain,* [b]*University of Alicante, Alicante, Spain*

1. Introduction. The promise of smart cities

> *The city, in addition to a built architectural ensemble, is the relational space where people live and produce society, where they are closer to the rights and duties of citizens*
>
> (Puente Burgos, Peñaloza Quintero, & de Gracia-Soriano, 2019).

An increasing number of local governments want their municipalities to have the smart city label. However, specialized literature recalls how the concept of smart city has varied in meaning over time and from different perspectives (Hollands, 2008), making it difficult to provide a single definition, and therefore the development of a smart city becomes complex. In this reflective work, it was decided to start from the definition provided by Tomàs and Cegarra (2016, p. 47), who conceive the smart city as a "more efficient, sustainable, participatory city with a higher quality of life for their inhabitants thanks to the use of Information and Communication Technologies (ICT)." These authors affirm that the concepts of participation, sustainability, and efficiency are at the core of the smart city paradigm. In this way, the unique association between smart cities and technology is broken to give rise to a local political management system not without ideology (Visvizi & Lytras, 2018, 2019a, 2019b; Winner, 1980).

The commitment to converting an urban space into a smart city is a political commitment, which involves discussion and decision of a more social or technological orientation, to choose between a public leadership or the distribution of responsibilities with private entities, among other examples. For this reason, the development of each smart city has its own specificities. There are as many smart city models as there are regions that are committed to this type of cities. It is in this context that sociology takes on special relevance as a fundamental discipline to apply to the study of smart cities. By observing smart cities through the lens of sociology, it is possible to identify the actors that build the smart city

Smart Cities and the UN SDGs. https://doi.org/10.1016/B978-0-323-85151-0.00013-0

and evaluate the role of citizens in this process (March & Ribera-Fumaz, 2016), discovering and describing the meanings, sensations, problems, potentialities, etc., in short, the multiple realities and experiences that people have in and with the city and its context.

The development of smart cities is understood within the framework of the evolution of the city itself. Especially, it is based on two trends characteristic of contemporary society: the growing urbanization and the digital revolution (PwC-IEBS, 2015). In 2008, the UN calculated that, for the first time in the history of the planet, there were more people living in cities than in rural areas. In this last decade, data and projections indicate that the trend toward population concentration is and will be increasingly important: in 2018, in its annual report on the distribution of the world population in urban centers, the UN estimated that 55.3% of the inhabitants of the planet lived in a city and that in 2050 it would reach 66%.

The landscape of an urbanized world supposes not only a concentration in demographic terms but also in economic terms. More than 80% of world GDP is generated in the 600 largest cities on the planet (McKinsey Global Institute, 2011). In other words, cities are the nerve center of economic activity. To the increase in the population of cities worldwide, the unstoppable process of digitization and technological connection of spaces is added. According to the latest estimates of the International Telecommunication Union (ITU), in 2019 53.6% of the world population (approximately 4.1 billion people) were Internet users. The ITU warns that the average year-on-year growth rate since 2015 (with 17% of the population connected to the Internet) has been 10%, although it asserts that in recent years this growth has slowed down because in some parts of the world saturation levels have been reached. The Internet is the basis for digital transformation, that is, the Internet is essential for the integration of digital technology anywhere on the planet.

The challenges that cities face are reflected in Resolution A/RES/70/1 of the United Nations General Assembly, approved on September 25, 2015. It develops the 2030 Sustainable Development Agenda, in which includes the 17 Sustainable Development Goals (SDGs) made up of 169 targets and 230 indicators that must be implemented, monitored and evaluated in this time frame, with the necessary commitment from the international community as a whole. All this seems to point to the need to create cities of the future that are capable of guaranteeing spaces of well-being and quality of life for citizens in their place of residence. Thus, cities as spaces for human and economic development are destined to develop a multitude of processes, which seem to lead us toward the long-awaited future for decades, where technology becomes solely at the service of citizens to improve their lives. All kinds of spaces and organizations undergo a multitude of key processes such as information technology, cloud computing, artificial intelligence technologies, augmented reality, virtual reality, social networks or the internet of things (Pérez-del Hoyo & Mora, 2019).

The information referred to so far suggests that smart cities grow under the promise of an efficient society based on the use of sensors, fixed and mobile, as well as the technologies derived from them. The historical attention around these ideas and facts shows that this bet has as its start and push engine, technological development based on the democratization of certain resources such as mobile applications, open-source systems based on Arduino and the growing expansion of the Do It Yourself (DIY) culture.

The development of smart cities does not occur equally in all cities, as indicated above. The research challenge posed by this analysis is to determine what the common and specific elements in the different cases are who are the actors that build the smart city, what role do citizens play in its construction, what governance models are developed, what type of policies are carried out and with what objectives (March & Ribera-Fumaz, 2016; Tomàs & Cegarra, 2016). In the details, in what is hidden most of the time, is where the "key" is found to think and rethink smart cities conceived as smart cities, as healthy cities, while promoting the achievement of the SDGs defined by the UN.

The rest of this chapter is structured as follows: Firstly, "The role of sociology in the construction of smart cities" is an introduction to the necessary and timely alliance that should exist between this discipline and the processes of construction and definition of this type of city. It is, therefore, a reflective section that is expected to serve as an invitation to work in this direction, both from research and from innovation, development and evaluation of this type of projects, especially when it is expected that smart cities are in harmony with the SDGs, of which there is no doubt about their high social and human component, but also technical, environmental, political and economic, and therefore, sociological.

Next, in the section "Human action and social participation in smart cities" a meta-analysis of the authors' research career is presented, exposing the importance of action research in the set of processes that intervene in a social organization such as a city while offering how this type of methodology can actively contribute to the development of smart cities and achieve the majority of SDGs collected by the UN.

Thirdly, "The relevance of time policies for the development of smart cities within the framework of the SDGs" exposes the need to understand the experiences in the urban environment of the population from a not only spatial but also temporal perspective. And not only from an economistic but also social and human vision, concerned with structural welfare and well-being, and therefore closely related to the SDGs focused on social equality and inequalities.

It is necessary to point out that this chapter is presented from a scientific, technical, and structural perspective in which sociological arguments and strategies are offered from the academic and professional experience of the authors, which are understood to be adequate and valid for the integration of the elements that give meaning to this book: SDGs and smart cities.

2. The role of sociology in the construction of smart cities

It is likely that when you think of a city you conjure up images of an architectural ensemble in your mind. However, it is less likely that when thinking about a city we evoke images that refer to the relational space where people live, coexist, and create a productive society where the rights and duties of citizenship are concretized and effective, or not (Cordoni, 1993).

For sociology, the city is a dynamic concept that refers to the relationships and spatiotemporal coordinates in which human beings produce senses and activities that give rise to lifestyles, models of production and consumption, in relation to the environment, models of social organization. Of course, the most obvious physical element is perhaps the architectural one, but it is also in the city where different sociocultural and ethical norms are configured and produced. In this sense, the city is, *grosso modo*, a physical and social space in transformation aimed to achieve a more democratic model, requires a conception of citizen participation where equality, cooperation, and responsibility of the different actors are present.

Because of this, the complexity of an analysis of a territory that seeks to become an intelligent environment requires the consideration of physical, environmental, and landscape aspects as well as social, anthropological, cultural, historical, architectural, urban, economic, and public health dimensions, among others. This is one of the reasons why sociology must be present in the design and evaluation process of a smart city since its historical, academic, and scientific trajectory is nourished by knowledge that transcends the classic boundaries between disciplines. Therefore, sociology is capable of efficiently achieving the integration of the different objectives that may arise in the development of a smart city related to the Sustainable Development Goals (SDGs), many of which have been working academically from this discipline for decades.

Each science and each branch of knowledge has its own object of study. Sometimes, as in the cases of formal and natural sciences, they are understood as independent of the social being. However, it cannot forget that, although their object of study is neither human nor social, the productions of knowledge are built by people, and therefore, even in these sciences there is a social component. Although the object of investigation is not directly related to the social being, it is indirectly so since the person who investigates is a social being and cannot get rid of the said essence. As an example, a genius like Albert Einstein, who made great contributions to the understanding of the physical world and also contributed to the invention of the atomic bomb. Artifacts, although they seem to have an independent character, are not devoid of ideology (Winner, 1980). They are not devoid of human and sociocultural properties and characteristics.

That is why sociology is necessary both to design a smart city connected with the population that passes through it, and to serve as a supervisory discipline of human action, regardless of technological, economic, and political guidelines. In summary, the design and technical-methodological implementation of a smart city must achieve an optimal level of pragmatism (Lytras & Visvizi, 2018), methodological caution and alertness, as well as an understanding of the uses and needs of the individual and collective population.

3. Human action and social participation in smart cities

> *Cities play a fundamental role as engines of the economy, as places of connectivity, creativity and innovation and as service centers for their surroundings. However, they are also places where problems such as unemployment, segregation or poverty are concentrated. That is why, both in the design and in the implementation of the different actions that may occur in the place in which one lives, citizen participation and, therefore, the different social agents must be participants in the planning of the different projects that can be developed*
>
> (Jareño-Ruiz, de Gracia-Soriano, & Jiménez-Delgado, 2018: 238).

With the development of smart cities, human action and social participation acquire controversial positions. Political-economic discourses and action seem to be focused on the citizen, but it is usual that these actions are not the product of citizenship or the result of a deliberation process by at least the residents of the city. Most local governments that bet on projects for the development of smart cities have technology companies as partners, so it is not difficult to imagine the importance of the role that the technology sector plays in these urban spaces. Certain authors affirm, after analyzing cases from smart cities such as Singapore or Barcelona, that there has been a growth in direct relations between local governments and companies, while this process has not followed the same trend with civil society. In other words, they consider that citizens have been left out of the definition of a smart city, even though they are paradoxically the recipient of its potential benefits (Tomàs & Cegarra, 2016).

These benefits include, theoretically, a certain improvement in the communication between the group of citizens and their leaders, provided in practice by consultative applications (John, 2001) that were elaborated seeking a horizontal, flexible, and open urban governance, a concept understood as "a process of coordination of actors, of social groups, of institutions to achieve collectively debated and defined objectives" (Le Galès, 2004). It is the idea of involving not only the general administrations of the different states but also municipalities, regions, the private sector, and civil society (GV, 2019; UN, 2015). Therefore, if smart cities are mostly based on these principles, it is necessary to ask what is not working for efficient social participation in the construction of cities.

Based on the work of Kübler and Wälti (2001), urban governance in smart city projects has two different interpretations. On the one hand, the population has greater power in decision-making and in negotiation processes when compared to traditional government structures—which in the best of cases, informed citizens of the decisions made and the actions taken. But the creation of regulations for citizen participation and the instruments chosen at random, in many cases, do not guarantee real participation. It is on this other side that a pessimistic vision emerges in which technology-based companies and other economic stakeholders invade these spaces of deliberative democracy to achieve their particular objectives. Although the smart cities discourse focuses on efficiency and quality of life, these objectives are at the service of a model of an entrepreneurial city, according to the work carried out by Harvey (1989). For some authors, the smart city is a high-tech version of the entrepreneurial city, where the main economic benefit is sought, which implies a greater influence of private sector actors and a "letting down" of civil society in the construction of the model and evolution of your city (Hollands, 2008).

A possible scenario in the approach of a smart city would be one in which the decision-making processes are conditioned, if not defined, by the results of certain sensors designed by a group of people, more or less experts, who build their designs from mainly theoretical components and with a low component of social participation, in other words, from a fundamentally technical point of view. In this scenario, it is not difficult to imagine a city in which the population that lives and transits through it stops making certain decisions because they are made by these sensor systems, programmed with a certain logic. In any case, this logic is a key aspect to take into account since it will define one or another city model, which could have as a consequence, a population that stops making certain decisions about what is incumbent on it.

To avoid this, citizen participation can be useful. It is probable that some actors involved may consider that citizen participation poses a certain risk for the development of any decision (let us think that participatory processes could be solved, for example, in the discard of smart cities) and that is why it is necessary to ask about of the type of city that you want to rebuild and in which you want to live. As in many other aspects of societal life, it is important to actively ask yourself the why and for what of things that are done.

Implementing a smart city in an environment where its population does not want such a project can be considered a problem. If the economic-political commitment is for this urban environment and operation model, then the population should be consulted and participate in the decision-making processes of that project, assuming the complexity and possible risks of this strategy, as well as the absence of it. In any case, any entity should seriously ask about the right to privacy of citizens, their opinion, who is likely to be benefited and who is likely to be harmed by this model (Jareño-Ruiz, de Gracia-Soriano, & Jiménez-Delgado, 2019). Should a smart city be inclusive and open to the changes and dynamics of its population and society? Or does it require some exclusion and a rigid society? Can it contribute to the reduction of certain inequalities? Or will it contribute to their increase?

One of the pillars on which the smart city phenomenon seems to be based is the idea of a more efficient and sustainable environment due to intelligently programmed decision-making processes, such as traffic management, lighting, pollution, parking, public services, irrigation systems, and accessibility. Although it is easy to see the potential of a scheme like this, one must also take into account the risk of controlling human action, regardless of people.

As can be glimpsed, a smart city has a strong technological component, but it remains lame and becomes a dangerous project if it does not also have a strong participatory sociological component. That is why a smart city, in the set of scenarios proposed, requires multidisciplinary teams, although desirably transdisciplinary, to tackle these projects by adjusting them to the societal reality in which they will be immersed.

4. The relevance of time policies for the development of smart cities within the framework of the SDGs

One of the probably strongest applications of smart cities is framed around urban mobility. In this context, an aspect that is not usually taken into account is the one that refers to the social uses of time that people apply to different moments and scenarios: life, work, and city times (Torns Martín, 2011). These new categories of social time, as a general concept, open doors to new perspectives, problems and causes of inequality that affect the foundations of an environment of welfare, *lato* sensu, and well-being (de Gracia-Soriano, 2019).

City time is a concept coined by Teresa Torns to refer to the styles and rhythms of life that people develop in the urban environment. It not only refers to the styles and rhythms of life of citizens but also to those other population profiles that make use of this environment, for example, workers or tourists. Reflecting and investigating about the times of the city is especially interesting for any municipality because it allows knowing what people's experiences are in the urban environment, from a spatiotemporal perspective. Probably one of the most widespread types of research in this context is mobility studies, which lead to certain public policies. However, they are not the only type of research that is currently being applied within the framework of city times and public policies:

> *Despite their heterogeneity, two large groups of city time policies seem to be drawn. In the first place, urban time policies or policies that consider the city as a material space in which to plan, regulate and order the multiple social uses of time in the urban territory. Secondly, time policies that are aimed at facilitating the well-being of citizens and that also take everyday life as a stage for action [...] Time policies should begin by questioning the hegemonic centrality of time of paid work (read daily workday) in the current socio-economic organization and in the daily life of people*
>
> (Torns Martín, 2011).

Understanding mobility from a uniquely economistic conception involves generating urban structures only destined for this area and also a rigid conception of urban life that ends up generating totalizing temporal structures (Rosa, 2016). However, people reside and use urban space for many other purposes, and therefore a comprehensive strategy cannot ignore this fact. Following the lines drawn by Torns Martín and de-Gracia-Soriano, a smart city must contemplate the diversity of times, spaces, rhythms and lifestyles that occur in it. In other words, a smart city must be open and attentive to societal complexity and contribute to the well-being of citizens through the investigation of their experiences of time, particularly those aspects that become stressors and generate clinical symptoms in people.

However, the analysis of reality from temporal perspectives is not usual, despite its daily nature, or precisely for this reason. In this sense, the development of smart cities has the opportunity to integrate temporal analysis in their designs from the occupied, liberated, and transitional times (de Gracia-Soriano, 2019) and the three spheres defined by Teresa Torns (Fig. 1).

"The organicist vision of the Smart city considers the urban space as a system of systems, as a flow of data that must be monitored and optimized" (Tomàs & Cegarra, 2016, p. 47). Along these lines, it is worth considering the desirability that smart cities contemplate in their fundamental structure, a Sociological Observatory of the temporary uses and experiences of the population. This permanent observatory would allow both the improvement of municipal management and the production of relevant data and information for all citizens and other actors.

As it has been reflected in this chapter, that local and regional governments know what the social realities of their municipalities or regions are, is an aspect of great relevance for any context that wants to have the label of "intelligent." One point in which this review aims to emphasize is the close relationship that must exist between a smart city, the population, and a healthy environment, both mental and physical. In other words, the joint integration and development of several of the SDGs defined in the 2030 Agenda becomes necessary. The following set serves as an example (UN, 2015):

- Guarantee a healthy life and promote the well-being of all at all ages (SDG3—Good Health and Well-being).

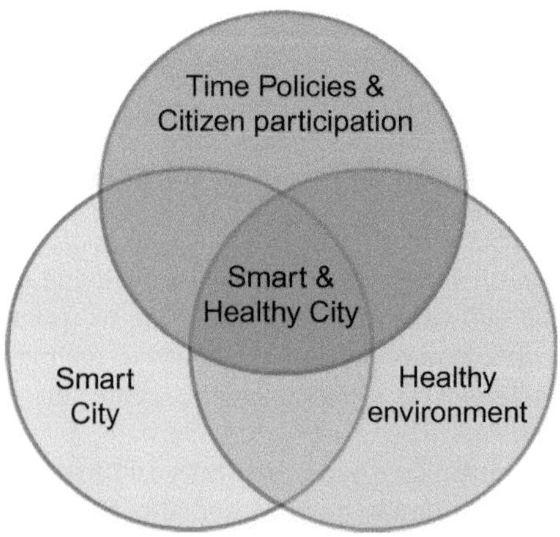

FIG. 1 Scheme for a healthy and smart city. *(Source: Own elaboration.)*

- Promote sustained, inclusive and sustainable economic growth, full and productive employment and decent work for all (SDG8—Decent Work and Economic Growth).
- Build resilient infrastructure, promote inclusive and sustainable industrialization and foster innovation (SDG9—Industry, Innovation, and Infrastructure).
- Reduce inequality within and between countries (SDG10—Reduced Inequalities).
- Make cities and human settlements inclusive, safe, resilient, and sustainable (SDG11—Sustainable Cities and Communities).
- Guarantee sustainable consumption and production patterns (SDG12—Responsible Production and Consumption).
- Promote peaceful and inclusive societies for sustainable development, facilitate access to justice for all and create effective, responsible, and inclusive institutions at all levels (SDG16—Peace, Justice, and Strong Institutions).

5. Discussion

This chapter shows some of the debates and key aspects of the construction and development of smart cities from a sociological perspective. The research experience of its authors suggests that a smart city will not be able to reach its minimum state of effectiveness if transdisciplinary teams are not incorporated in the design, evaluation, and monitoring processes that allow the correct integration of knowledge about urban life. Otherwise, it seems that the most probable scenario is that of facing certain risks that endanger coexistence and therefore the daily well-being of the populations that live and transit through these municipalities.

The development of cities as smart cities, taking into account the democratic frameworks of modern states, should not be imposed, but consensual. In this line, as has been detailed in the preceding pages, a smart city cannot be conceived the same as another one, even if they use the same technologies. In other words, the idea of a smart city cannot be understood homogeneously, despite the homogenizing process of globalization. Quite the contrary, for a smart city to have a relative degree of success over time, it must adapt to the demands of the people, and to the plurality of rhythms of its permanent residents, temporary residents, and visitors, taking into account the diverse objectives that each one of these segments of the population can have.

That is why it can not only serve economic interests but also social and community interests, where in addition to the value for the quantitative, attention is given to the qualitative and to the problems of each age and the more or less collective particularities of the people who live and/or will live in these places. In other words, the smart city cannot become a prison for present or future citizenship. Rather, it could be expected that it would provide a space of freedoms consistent with the ethics and morality of each era, without losing focus on individualities and their heterogeneous needs and preferences.

In these pages, it has been related how the implementation and use of mobile and fixed sensors in the urban environment poses an ethical problem in the first instance, especially through image-based sensors or those that can identify users or some personal data, for example, through Wi-Fi connections or the increasingly widespread mobile applications, or apps. Regarding the latter, there is no denying the bet that is being made from different areas to have a good reception among the public, as can be seen, for example, in the recent applications to control and manage the COVID-19 pandemic (see, e.g., the case of South Korea or its recent establishment in the Basque Country, Spain). However, despite the potential that these types of instruments may have, especially at a general and structural level, the fine line that separates them from the rights of individuals is extremely fine and would require a review of both these rights and ideas of the common good or of general interest. In other words, accepting this type of instrument is as much as saying that a system that uses them has a strong invasive character in people's lives and their fundamental rights.

Similarly, it seems clear that not all types of sensors and tools that can serve the development of a smart city are equally invasive nor do they all have the same scope. The turning point can be found precisely in the idea of those instruments that are capable of identifying specific users, or their profile features. Obviously, a sensor that reports parking availability does not have the same social and human reach as a sensor that identifies you as a carrier of the virus and alerts other people around you, or around the world. Therefore, an analysis of the convenience or not of a smart city in a specific region cannot ignore the possible consequences that it will have or could have on the population.

Among other reasons, this is one of the most important for which citizens, and in general, any segment of the population that can transit or use the space in which a smart city is located, must participate both in the design phase of the infrastructures as in local or regional decision-making processes.

Making use of the levels of participation defined by Alberich (2004) (information/training; consultation and discussion; and, participation in management/shared management), it is observed that the majority of municipalities bet on their public policies for models based on the first and second levels, being rare those that reach or bet on the third level. The instruments for citizen participation designed in smart cities projects are fundamentally based on information and consultation actions, leaving the first two levels incomplete because they need to be complemented with training in the first one, and with debate, in the second. Lack of attendance should not be a problem for the correct performance of deliberative activities. However, if training is necessary when the information is not understood, as well as if the results of the consultation are not debated, the potential of Participatory Action Research (PAR) techniques is lost. The dialog between the administration and the citizens must be permanent, but the spaces and times that make it possible must be enabled. If not, it will be impossible to advance to the third level, that is, to advance to the stage of real shared management, comanagement that allows citizen self-management.

As mentioned in the preceding pages, one of the strongest arguments for those who are in favor of the development of smart cities is that of sustainability (PwC-IEBS, 2015). In a research carried out in Spain to 1200 citizens and 40 experts, it was concluded that the population is especially interested and concerned about the applications that could be developed in a smart city, highlighting the fields of health, the environment, education, and citizen security. In this sense, it should be noted that these voices propose smart cities as a tool to face the new challenges derived from demographic growth through mechanisms that make resource consumption more efficient (from optimizing water resources to energy and the reduction of pollutants).

However, and returning to the question that has served as a guiding question in this chapter (are smart cities accessible for all people?), it is also necessary to reflect on the relevance of the development of isolated smart cities, that is, the emergence of models of a smart city in geographic spaces not technologically developed. The common elements that define smart cities are linked to technique, digitization, and the use of technology for the provision of urban services that improve the quality of life of its citizens, facilitating a new way of living in the city. This predominantly local dimension must be rethought since the cities are not isolated, but are part of more complex urban structures (conurbations, regions, metropolitan areas, etc.). It does not seem "smart" to amount smart cities that are not interconnected. The development of projects already started, like those that may be started, must be able to transform the relationship between local entities, companies, and citizens. Would the existence of a smart city in an undeveloped metropolitan area make sense? As the reader will be able to see, there are more questions than answers, and it is necessary to stop at them.

As a closing reflection of this chapter, the promise of smart cities referred to in the Introduction is collected. To do this, we emulate this idea from its homologous on the acceleration of Hartmut Rosa (2016), which is closely related given the technological-economic nature that surrounds smart cities. The unfulfilled promise of modernity is one in which the future has translated into a present with a greater degree of contingency in societies in which paradoxes are found that end up generating states of alienation in populations and individuals.

Similarly, it is necessary to conclude with the warning, from the present that will be past, that the promise of smart cities does not become another unfulfilled promise within the framework of the technology-economy union, a trap for the populations and municipalities that adapt to them. That the bets for these city models are not perverted or show that they have hidden interests behind the ideas of participation, sustainability, and efficiency.

This chapter has attempted to expose some elements that are considered necessary for the reflection, design, evaluation, and monitoring of smart cities. Some of them are currently active debates and others can be taken as prudent elements. In any case, despite the critical tone of the chapter, the authors recognize the potential of smart cities and have tried to express their contributions to a city model that extends its concept to that of smart and healthy cities, committed to everyday well-being of the population.

References

Alberich, T. (2004). *Easy guide to citizen participation. Management manual for the promotion of citizen participation in town halls and associations*. Madrid: Dykinson.

Cordoni, E. (1993). Las mujeres cambian los tiempos [Women change times]. *Cuaderno de Relaciones Laborales*, *2*, 221–227.

de Gracia-Soriano, P. (2019). Tiempos sociales y desigualdades: una aproximación [Social times and inequalities: An approximation]. In M. Jabbaz-Churba, J. A. Rodríguez-del-Pino, & N. Navajas-Pertegás (Eds.), *Miradas de género: una sociología sin barreras, cerraduras ni cerrojos [Gender views. A sociology without barriers, locks or bolts]* (pp. 47–56). Barcelona: Icaria Editorial.

GV. (2019). *The SDGs and the private sector: Opportunities for action in the Valencian community*. Valencia: Consellería de Transparencia, Responsabilidad Social, Participación y Cooperación – Generalitat Valenciana.

Harvey, D. (1989). From managerialism to entrepreneurialism: The transformation in urban governance in late capitalism. *Geographiska Annaler*, *71*(B), 3–17.

Hollands, R. (2008). Will the real smart city please stand up? Intelligent, progressive or entrepreneurial? *City*, *12*, 303–320.

Jareño-Ruiz, D., de Gracia-Soriano, P., & Jiménez-Delgado, A. (2018). The relevance of citizen participation in European strategies for sustainable and integrated urban development. A case study. In B. Tejero Claver, O. Bernad Cavero, & C. Lechuga Jiménez (Eds.), *Investigating cutting edge content* (pp. 237–250). Barcelona: Gedisa.

Jareño-Ruiz, D., de Gracia-Soriano, P., & Jiménez-Delgado, M. (2019). Citizen participation in natural and sociocultural landscape: Methodology and main results. In A. Jiménez-Delgado, & J. Lloret-Mauri (Eds.), *Health, wellbeing and sustainability in the Mediterranean city* (pp. 64–74). New York: Routledge, Taylor & Francis. https://doi.org/10.4324/9780429401572-6.

John, P. (2001). *Local governance in Western Europe*. London: Sage Publications.

Kübler, D., & Wälti, S. (2001). Metropolitan governance and democracy: how to evaluate new tendencies? In P. Mclaverty (Ed.), *Public participation and developments in community governance* (pp. 115–140). Ashgate: Aldershot.

Le Galès, P. (2004). Gouvernance. In L. Boussaget, S. Jacquot, & P. Ravinet (Eds.), *Dictionnaire des politiques publiques* (pp. 242–250). Paris: Presses de Science.

Lytras, M. D., & Visvizi, A. (2018). Who uses smart city services and what to make of it: Toward interdisciplinary smart cities research. *Sustainability*, *10*(6), 1998.

March, H., & Ribera-Fumaz, R. (2016). Smart contradictions: The politics of making Barcelona a self-sufficient city. *European Urban and Regional Studies*, *23*(4), 816–830. https://doi.org/10.1177/0969776414554488.

McKinsey Global Institute. (2011). *Mapping the economic power of cities*. Available from http://www.mckinsey.com/global-themes/urbanization/urban-world-mapping-the-economicpower-of-cities.

Pérez-del Hoyo, R., & Mora, H. (2019). Knowledge society technologies for smart cities development. In A. Visvizi, & M. Lytras (Eds.), *Smart cities: Issues and challenges: Mapping political, social and economic risks and threats* (pp. 185–198). Amsterdam: Elsevier.

Puente Burgos, C. A., Peñaloza Quintero, R. E., & de Gracia-Soriano, P. (2019). Exploring the (re) construction of healthy destinations in the Mediterranean. In A. Jiménez-Delgado, & J. Lloret Mauri (Eds.), *Health, wellbeing and sustainability in the Mediterranean city* (pp. 50–63). New York: Routledge, Taylor & Francis. https://doi.org/10.4324/9780429401572-6.

PwC-IEBS. (2015). *Smart cities: The digital transformation of cities*. PwC Public Sector Innovation Center, IE Business School, Telefónica. Available from https://m2m.telefonica.com/libroblanco-smart-cities/media/libro-blanco-smart-cities-esp2015.pdf.

Rosa, H. (2016). *Alienation and acceleration. Towards a critical theory of temporality in late modernity*. Buenos Aires: Katz.

Tomàs, M., & Cegarra, B. (2016). Actors and models of governance in smart cities. *URBS Revista de Estudios Urbanos y Ciencias Sociales*, *6*(2), 47–62.

Torns Martín, T. (2011). Reconciliation of work and family life or co-responsibility: The same discourse? *Revista Interdisciplinar de Estudios de Género*, (1), 5–13.

UN. (2015). *Resolution A/RES/70/1 approved by the General Assembly on September 25, 2015: Transforming our world: The 2030 agenda for sustainable development*. United Nations General Assembly (UNGA). Available from https://www.un.org/ga/search/view_doc.asp?symbol=A/RES/70/1&Lang=S.

Visvizi, A., & Lytras, M. D. (2018). Rescaling and refocusing smart cities research: From mega cities to smart villages. *Journal of Science and Technology Policy Management*, *9*(2), 134–145. https://doi.org/10.1108/JSTPM-02-2018-0020.

Visvizi, A., & Lytras, M. D. (2019a). Smart cities research and debate: What is in there? In A. Visvizi, & M. Lytras (Eds.), *Smart cities: Issues and challenges: Mapping political, social and economic risks and threats* (pp. 1–14). Amsterdam: Elsevier.

Visvizi, A., & Lytras, M. D. (2019b). Reflecting on oikos and agora in smart cities context: Concluding remarks. In A. Visvizi, & M. Lytras (Eds.), *Smart cities: Issues and challenges: Mapping political, social and economic risks and threats* (pp. 333–339). Amsterdam: Elsevier.

Winner, L. (1980). Do artifacts have politics? *Daedalus*, *109*(1), 121–136.

Chapter 14

Smart city as the city of knowledge

Robert Laurini[a,b], Giuseppe Las Casas[c], and Beniamino Murgante[c]
[a]*Knowledge Systems Institute, Skokie, IL, United States,* [b]*LIRIS, University of Lyon, Lyon, France,*
[c]*LISUT, School of Engineering, University of Basilicata, Potenza, Italy*

1. Introduction

In 2015, more than 190 world leaders have adopted 17 Sustainable Development Goals (SDGs) and 169 associated targets, defined by the United Nations (UN), to be achieved before 2030 (United Nations, 2015a). The main aim of these goals and targets is to stimulate fundamental actions for humanity and the planet before 2030.

Among these 17 goals, Goal 11 "Make cities inclusive, safe, resilient, and sustainable" has a direct connection with Smart Cities, another research field which has been debated in the last 10 years. A simultaneous analysis of Goal 11 and Smart Cities can lead to advantages of economies of scale, of complementarity of many aspects, and of avoiding unnecessary overlaps. A lot of experiences related to smart cities are mainly concentrated on technological aspects without considering if all these innovations are useful for the city (Murgante & Borruso, 2013, 2014). In order to accelerate national progress to achieve the sustainable development goals (SDGs), the Agenda 2030 for Sustainable Development (United Nations, 2015b) has been defined ensuring that no one will be left behind and moreover to reach the furthest behind first (Fleurbaey, 2018).

There are five key factors in analyzing who is left behind: discrimination, geography, governance, socioeconomic status, and vulnerability to shocks (United Nations Development Programme, 2018). In order to implement the Agenda 2030 these factors can be applied adopting an integrated approach based on three axes:

- Examining all kinds of weaknesses that people have to face in the five factors.
- Empowering people including those left behind in decision-making by means of inclusive mechanisms of civic engagement.

Smart Cities and the UN SDGs. https://doi.org/10.1016/B978-0-323-85151-0.00014-2

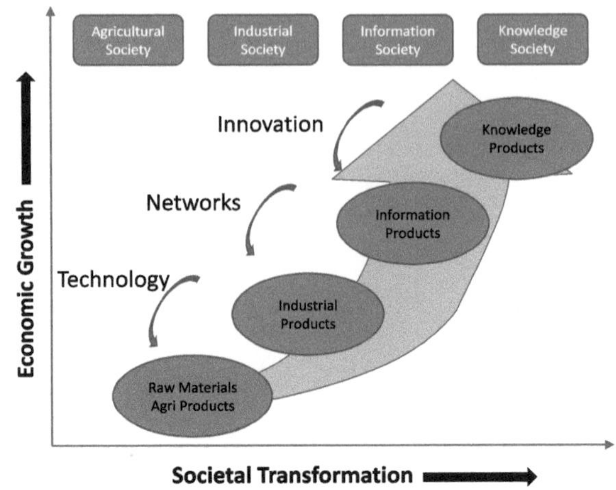

FIG. 1 Evolution of the civilization (Paul, 2014).

- Enacting change implementing policies, laws, reforms, interventions for reducing inequalities and for increasing the minimum standards of well-being.

In order to reach this result, it is fundamental to increase the level of knowledge in cites and societies.

Several authors (Bindé & Matsuura, 2005; Castelfranchi, 2007; Lytras & Sicilia, 2005) say that we are entering the knowledge society. For instance, Paul (2014) explains that gradually (Fig. 1) the civilization has passed from the agricultural, industrial, information society to the current knowledge society.

What exactly does anyone mean speaking of the knowledge society? And about cities, not only at management level but also for the daily life of citizens? Here are the research questions that will be examined in this chapter. The discussion in this chapter is structured as follows. First, various definitions of the knowledge society are examined with special attention to encyclopedism, governance in companies and in urban planning with respect to rationality. Do not forget that some knowledge chunks can come from public participation and overall electronic participation. This chapter develops also examples of encoding rules, as knowledge components, in urban planning.

In the conclusion, it will be stressed that new acquired knowledge demonstrates the idea of revisiting strategic approach for economic and spatial planning.

2. The knowledge society: Issues and implications

What do we mean exactly when speaking about the knowledge society? What could be the differences with the information society? In this section, we will try to answer those questions, by clarifying both concepts of information and knowledge. For computer scientists, it is common to distinguish data, information, and knowledge. In a very simple way, whereas data are bits, numbers, and strings of characters, information corresponds to data with their meaning, and knowledge is defined as information potentially useful to solve a problem. In English, the word "knowledge" is very general. When it is important to consider a piece of knowledge, several authors use the expression "knowledge chunks"; whereas several connected knowledge chunks can be regrouped into a "knowledge bundle," for instance, a set of knowledge chunks relative to freight transportation, water supply, touristic activities, etc. In fact, several categories of knowledge exist. Knowledge can be explicit or implicit; by explicit, one states that knowledge chunks can be explained to somebody else, perhaps with a natural language; by implicit, difficulties occur when explaining them.

From a computing point of view, two main sources of knowledge exist:

- From experts: in this case, experts can give a few sentences corresponding to what they know; those knowledge chunks are often explicit; in this category, we can easily add legal and physical laws and also best practices.
- From data or text mining or from big data analytics (Shekhar & Vold, 2020): it may be based on frequency analysis, knowledge chunks can be extracted; sometimes their meaning is mainstream, but often the causal explanation is not immediate; however, everybody has to be aware of possible spurious correlations; in this category, we can include the analysis of sensor data in real time.

Moreover, according to Afgan and Carvalho (2010), "the development of the Knowledge Society is focused on the following objectives:

- To inspire and enable individuals to develop their capability to the highest potential level throughout life, so they can grow intellectually, be well equipped for work, can contribute effectively to society and enjoy active personal fulfillment;
- To increase knowledge and understanding for their application at local, regional, national level;
- To play a major role in shaping a democratic, civilized and intellectual society;
- To learn, evaluate, assess and validate economic, environmental, social and technological advancement to produce benefits based on the knowledge society."

Based on these considerations concerning the whole society, what could be told regarding the impact of knowledge, and especially geographic knowledge for smart cities and territorial intelligence?

Of course, there are other definitions and points of view. Our goal is not to examine all of them, but rather to identify the main characteristics of the knowledge city.

3. The knowledge city

Currently, 50% of the world population is living in cities. Consequently, the knowledge city must be considered as a piece of the knowledge society. But this is not only a problem of scale, but rather different levels of preoccupation. According to an NSF workshop (Edwards et al., 2012), knowledge infrastructure must include at least the following: education, libraries, publishing industry, intellectual property, research, knowledge-based industry, and knowledge expertise. In this section, we will not develop all those aspects, but rather emphasize only on knowledge-based education, knowledge-based economy in cities. The special case of knowledge-based urban planning will be treated in the subsequent sections.

As we consider human evolution, one can see that continuously men and women have tried to transmit their knowledge to their kids, i.e., the ways to survive personally and collectively in different contexts. Gradually the so-called the hunter-gatherer nomadic tribes evolved to Neolithic agriculture and small settlements characterized by a progression of behavioral and cultural features and changes, including the use of wild and domestic crops and of domesticated animals and marked by the development of metallurgy, leading up to the Bronze Age and Iron Age and the creation of cities. More or less in parallel, scripture was invented as a way to transmit knowledge. And more recently to the industrial economy and the digital age with the advent of smart cities. Do not forget that the covid-19 crisis must force people to increase their medical knowledge, to adopt new behaviors, and to invent new relationships between people.

Along this evolution, parents have taught their children how to survive in jungles, in deserts, in seas, in mountains, and in cities. In the future, it will be perhaps on other planets. In other words, one has to consider a mixture of adaptation and evolution of knowledge along millennia. It is not the simple accumulation of knowledge, but, overall, the reconstruction and the reorganization of knowledge may be done by abandoning some bundles of knowledge which were no more in adequation with the context.

In other words, the human condition is characterized by its ability to learn and transmit knowledge to other people. But, now with the acceleration of history, education must be revisited not only based on pure learning for the lifetime but also based on learning to learn.

Discussing with elder people, maybe with centenarians, they can explain that they were born in a different world and all their life they were obliged to learn and adapt to new contexts. And those who had not this potentiality have been and will be in great difficulties.

Presently, is education based either on information or on knowledge? As the goal is to learn something, it can be considered as information, the best knowledge in education is to learn how to learn. So, this is a set of methodologies which must be created, perhaps rediscovered, however, formalized in order to be easily understood, learnt, and transmitted to other people.

Throughout history, humans have invented a lot, and those skills, know-hows, and knowledge have been transmitted to other people; think about agriculture, metallurgy, scripture, etc.

During centuries, perhaps millennia, libraries were the place where to accumulate knowledge as, for instance, the famous Great Library of Alexandria which was said to store many hundreds of thousands of scrolls. Later in the 18th century, with many contributors, Diderot (1751) launched the huge project of the "Encyclopedia, or a Systematic Dictionary of the Sciences, Arts, and Crafts" which was supposed to concentrate the whole knowledge and know-how of this period. Fig. 2 gives the visual organization of knowledge back of the Encyclopedia, and with actual vocabulary, it can be seen as a preliminary of a global ontology.

Later, there were the inceptions of Encyclopedia Britannica, and more recently of Wikipedia with similar goals, but at different times and different technologies. Maybe, in the future, since knowledge and technologies are evolving, with other media, other trials will be made to organize knowledge.

However, knowledge-based education is supposed to be organized at the level of the world, maybe with some cultural or linguistic differences. However, back to cities, a sort of scale level could be distinguished, for example, in history and geographic, a greater focus could be made so that local citizens may have more information and knowledge about the place they live by, finally, more general and synthesized knowledge at upper level, and more detailed at lower level.

Back to education in smart cities, the question is: how to organize teaching so that knowledge can be easily accumulated, stored, used, and transmitted? For the moment being, several recommendations can be given

- Educate kids to the actual world and possibly to the future world.
- Increase their ability to learn by themselves by means of different media including common books and Internet.
- Educate for sustainable development.
- Educate grown-ups to learn on to increase not only their cultural level but also new technological, medical, ecological, and sociological issues.
- Educate to help people with physical, intellectual, and social disabilities.
- Educate to respect other opinions and to empower citizens especially regarding the life of their city.
- Educate to think about the discovery and the defense of general interest, at short term as well at long term.

3.1 Knowledge-based economic activities

Two kinds of companies must be distinguished, conventional businesses and consulting firms. Indeed, the production of consulting firms is essentially knowledge whereas others can be anything, from products to services (Yigitcanlar, 2015). Remember that knowledge can be defined as information potentially useful to solve a problem, whereas it is clear when facing a present problem, what kind of information could be useful for the future? Theoretically, any!

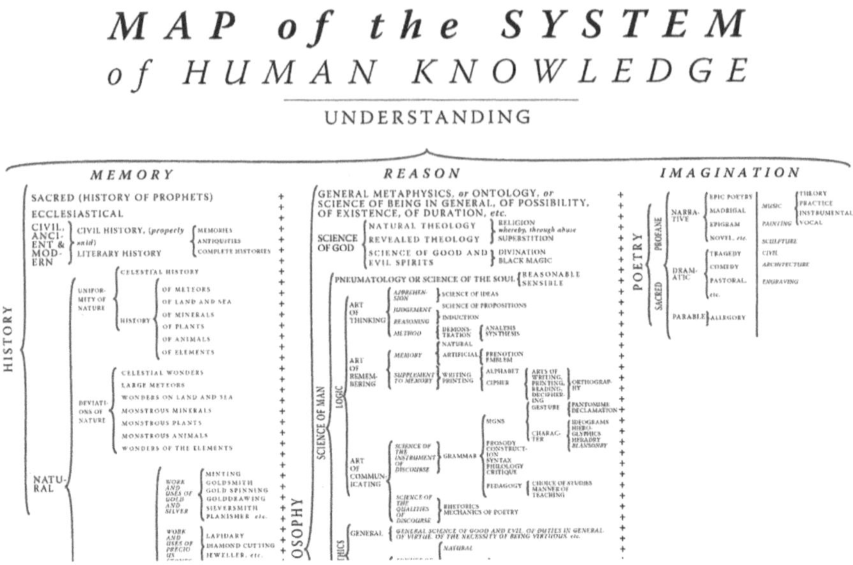

FIG. 2 Excerpt of the map of system of human knowledge within the Encyclopedia with three main branches: memory, reason, and imagination (Diderot, 1751). A complete original in English can be seen at https://quod.lib.umich.edu/d/did/tree.html.

3.1.1 Conventional companies

What kinds of knowledge companies use? Even if the knowledge vocabulary is little used, a lot of information coming from various origins will be the base of a company's knowledge. As illustrated in Fig. 3, knowledge can come from organizational theories, juridical context, guild recommendations, financial information, local authorities' regulations and projects, without forgetting information about customers, suppliers, competitors, and trade unions. Sometimes, when necessary, some of them call consulting firms for advice, to increase their efficiency. In addition, some companies must follow sociological trends and base their strategy on local authority projects, national information, and international opportunities. Finally, good knowledge management gives the company a competitive edge.

However, among knowledge chunks, one is very important, namely insider offences: in other words, it means that sometimes even if a company is aware of something, maybe financial information, it may not use it because it could be a criminal offense.

Second, in output (Fig. 4), we can mention knowledge bundles for internal organizations and regulations, production and human resources management, etc.

3.1.2 The special case of consulting firms

Differently from other companies, the essential outputs of consulting firms are the advice they sell to other companies.

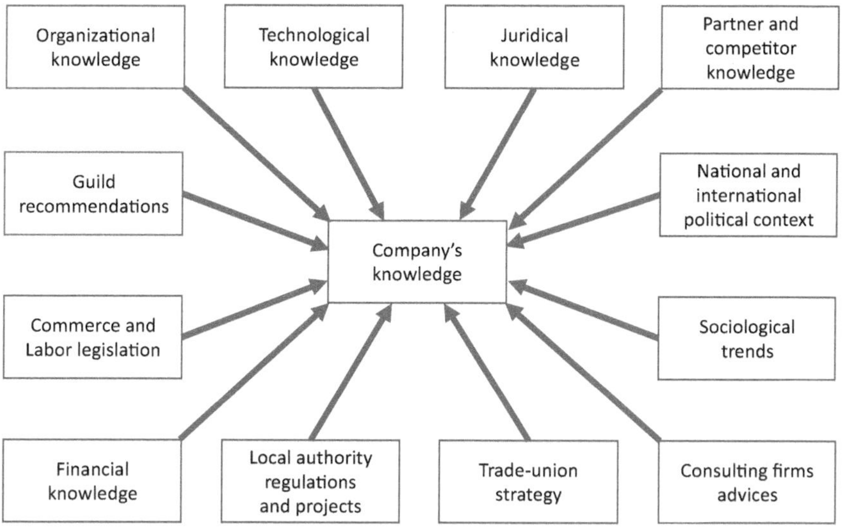

FIG. 3 Origin of company's knowledge.

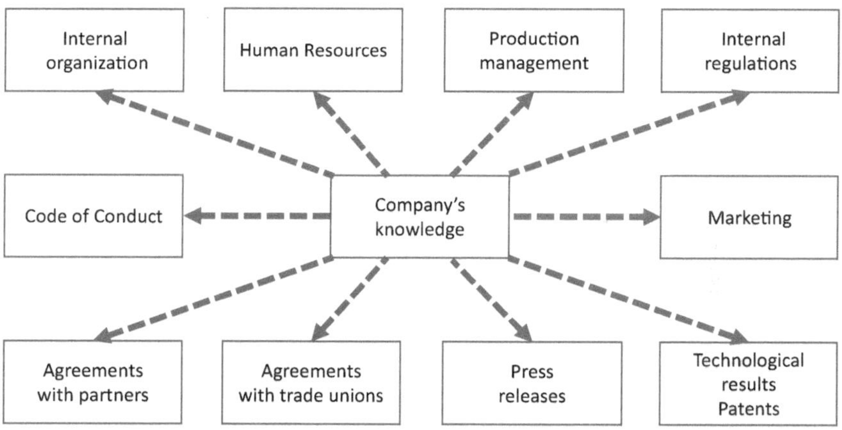

FIG. 4 Knowledge bundles issued by a company.

To illustrate their importance, according to Forbes (Valet, 2020), the consulting business boomed, growing 3.4% to revenues of $259 billion from more than 774,100 firms since 2014. In other words, more and more companies desiring to better their development strategy are seeking to ask the assistance of the knowledge business.

Another way to get knowledge is identified as big data analytics. Daily companies are creating billions of data and information pieces, so the goal is try to extract useful information and knowledge, first by data mining and then with deep learning.

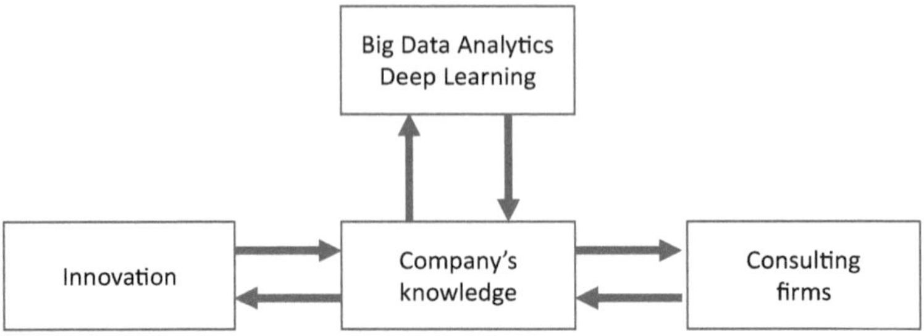

FIG. 5 Information and knowledge synergies.

Taking all that into account, cross-fertilizations can be discovered as depicted in Fig. 5: not only big data analytics, deep learning, innovation, and consulting firms bring fresh knowledge to companies but also this newly acquired knowledge can renovate visions.

In addition to that, nonprofit and not-for-profit organizations can contribute to the development of a city; and for them, knowledge is a key issue not only for their development but also for the whole economy. Among them, let us mention sportive, cultural, humanitarian, religious activities, etc.

4. Knowledge-based urban management and planning

In order to plan and manage a city, it is interesting to store and position data, information, and knowledge. As data are dealt with geographic information systems and information through metadata linked to GIS, currently there are no commercialized tools to deal with geographic knowledge, or more exactly urban knowledge. What could be the main characteristics of an urban knowledge repository, and how can it be used to plan and manage a smart city?

Even if urban and regional planning imply a vision only for a particular portion of the world, it is true that human settlements are representatives of a large portion of the outcome of human activities and of the human ability to change the Planet. So, any plan targeting urban transformation could mean to select objectives staying precisely tuned with a general view on the Planet.

First under an operational point of view, knowledge has to be based on information which corresponds to questions useful to identify problems, their location and people concerned and consequent objectives.

In this way, according to Laurini (2017), geographic and urban knowledge can be used for explaining, managing, monitoring, planning, understanding the past and innovating (see also Angelidou, Gountaras, & Tarani, 2012; Angelidou & Mora, 2019).

In the next section, after a short overview on the UN 2030 Agenda, we wish to stress the path from data to knowledge, starting from the need to know problems until to try to solve them by a plan activity. Along this path, we will develop competence and needs for education.

4.1 Objectives and problems

Following the large reflection on the geographic knowledge definitions, in the field of urban planning and management, it is interesting to analyze classic literature. More precisely, considering the cornerstones of strategic planning, a lot of authors (Chadwick, 1978; McLoughlin, 1969) have discussed the close relationships between objectives and planned actions and the application of the well-known principle "survey before plan" (Faludi, 1987; Geddes, 1915).

Using a temporal scale, we can

(a) Explain present and future trends, and understand the past trends.
(b) Monitor present trends.
(c) Manage (manage and monitor) present and future trends.
(d) Plan for solving problems of the future, through innovation.

In this general view, we are looking for those main contents needed to permit us to go from data to knowledge and creativity.

We think that a point remains central in the past as well as for the present or the future understanding: the most relevant and diffused social concerns.

In this short paper, it is not a question to present a list of concerns, but to reflect on a process where it must be clear that the scope of investigation is to support a collective decision.

An attempt to represent that process is made Table 1.

As the basis of strategic planning suggests, it is a process organized around an objective-based system and addressed toward a generalized consensus and implementation of the aftermath.

So, following Chadwick (1978), the objective detection remains very crucial; equally the central question remains: if the aim of the plan is to achieve those objectives, then how are objectives and social concerns linked?

Following some authors (Faludi, 1987) as objectives are to remove problems, and problems are everything that impedes that goal, the question becomes: if problem detection precedes objective detection or vice versa [first the chicken or first the egg? "Ovum ne prius extiterit an gallina?"]. Following classic, ancient Greek philosopher Parmenides showed how this famous paradox plays a role in explaining the relationship among *being* and *becoming* and, in our operational view, put the question: **problem first or objective first**?

This question that seems to be a very philosophical one is related to an attempt we are caring: to ground planning activity on a strong rational base (Las Casas, Scorza, & Murgante, 2019) in which any preconized action is proposed to solve some problems and if possible, by a multicriteria comparison, try to demonstrate if any proposed action is the **most efficient, equitable, and sustainable**.

TABLE 1 Actions of evolution from competence toward knowledge.

Actions **and connected knowledge level** → ↓ **Point of view**	Evaluate	Compare	Select and Propose
	To make a comparison with a measure unit	Through space and time to detect critical situations or alternative solutions	Choice plan or management solution and select; innovate when needed
	⇒ Competence	**⇒ Information**	**⇒ Knowledge and Creativity**
EQUITY	Explanation and an analytic description	List of problems and connected objectives and their evolution	“A planned course of actions” and the effects evolution
	DISPARITIES AND EQUITY	DISPARITIES AND EQUITY	DISPARITIES AND EQUITY
EFFICIENCY	Explanation and an analytic description	List of problems and connected objectives and their evolution	“A planned course of actions” and the effects evolution”
	POINT OF VIEW OF EFFICIENCY	POINT OF VIEW OF EFFICIENCY	POINT OF VIEW OF EFFICIENCY
SUSTAINABILITY	Explanation and an analytic description	List of problems and connected objectives and their evolution	“A planned course of actions” and the effects evolution”
	ON NATURAL RESOURCES CONSERVATION	ON NATURAL RESOURCES CONSERVATION	ON NATURAL RESOURCES CONSERVATION

4.2 Links with the UN sustainable development goals

So, the next reflection deals with the universality of those objectives. There is a link with juridical principles in occidental democracy where a rational and equitable use of resources is recognized as a general right. In France, as in Italy, e.g., emerge principles of equality and justice. The first article of the French Constitution states that "It ensures equality before the law of all citizens regardless of origin, race or religion." French Environment Charter (2004), which has validity as a constitutional principle, states that "in order to ensure sustainable development, choices to meet the needs of the present must not compromise the ability of future generations and other peoples to meet their own needs."

The same happen in Italy (Article 3): "all citizens are equal before the law." More in Article 9, the Italian Constitution puts under the responsibility of the Nation "the protection of the landscape and the historical and artistic heritage of the nation." And in Article 44: "In order to achieve rational land use and to establish fair social relations, the law imposes obligations and constraints on private land ownership, sets limits to its extension according to the regions and agricultural areas, promotes and requires land reclamation."

The same attention is reserved as the basis of the majority of European countries. For instance, we can quote the European Constitution in which the Treaty of 2004 states the sustainability and rationality in the use of resources and the need to guarantee citizens facing the Human Rights equality.

Briefly the 17 SDGs resume a long evolution developed in the entire world under the UN guidance which has produced a lot of documents, among other we underline "Guideline for urban and regional planning" where concepts are enlarged until they become the opportunity for deep innovations in practices.

So, the three basic principles are suggested in this work as a diligent selection, after a verification, that there is no contradiction and we can find a substantial accordance.

We can suggest that 17 SDGs are mainly addressed to equality and sustainability, but their very rich argumentation shows efficiency, i.e., rational use of any resource could be included in that general picture.

In effect, 17 SDGs seem to pertain to three groups: inequalities reduction (G1-G5, G10); sustainability and Planet and natural process defense (G6-G9 and G11-G15); peace and social justice (G16-G17) conclude the list as the most important precondition to realize remaining goals.

Once the robustness of that principle is accepted, in the following, we can try to show as they became operational in the knowledge process development. Indeed, as all those goals are defined for the whole Planet or society, they have all an incidence over cities. Especially, Goal 11 states that "Make cities and human settlements inclusive, safe, resilient and sustainable" and Goal 6 "Ensure availability and sustainable management of water and sanitation for all." In other words, this is the role of smart city governance to include those goals not only in the management of their daily activities but also for planning the future.

In other terms, those 17 SDGs must be transformed into knowledge chunks. Don't forget they belong to a "Knowledge Platform."

4.3 Equity, efficiency, and sustainability

(a) Competence

Under **equity**, we have to **evaluate** whether geographic or temporal situations are also different from each other. Competence, defined as techniques of measure or statistical indicators of inequalities, can become the object of a specific education (e.g., spatial distribution of health structures per capita or sick men percentage, statistical distribution of the distance among hospitals and citizens, etc.).

From the **efficiency** point of view, both absolute and comparative costs need to be taken into account (e.g., transportation or accessibility cost, construction and maintenance, running costs, etc.).

From the **sustainability** point of view, the common concern becomes the word conservation in its different aspects. Considering urban and regional planning, the main aspects with regard to human settlements and natural resource consumption and usual indicators are how much persons live in or in proximity of rare resources or use or abuse those? Which resources are under the risk of extreme consumption?

The output is a sort of explanation or an analytic description of a given situation.

(b) Information

The application of this kind of competence could generate **information** when a **comparative** exercise, aimed to put in evidence the most weak or critical situations, is implemented.

Changes through time is an aspect of **historical analysis**; it could be **creative** when important interpretations which **innovate** traditional interpretations are proposed; moreover, monitoring of it is self-evident.

The outputs are a list of problems and consequent objectives. The reached information is based on a relevant social interaction where people participate in the detection of critical situations comparing its needs and aspirations.

(c) Knowledge

In this exercise another level of creativity, maybe the most important, is that level where a course of actions is planned to transform a given situation.

Also, very important is the capacity to learn lessons from the past (historic point of view and the experience value) and also from successful experimentations made in other cities.

The output is a plan, i.e., "a planned course of actions" in which every single action can be justified because of its contribution to solve any diffused problem.

4.4 The importance of models in using data

During the past decades, the main problem in decision processes was the lack of data availability, but in recent years the wide diffusion of electronic devices containing geo-referenced information generated a great production of spatial data. Volunteered geographic information activities, public initiatives (e.g.,

Open data, Spatial Data Infrastructures, Geo-portals), Google and other spatial platforms, and also social networks produced an overabundance of spatial data, which, in many cases, does not help the efficiency of decision processes. The increase of geographic data availability has not been fully coupled by an increase of knowledge to support spatial decisions.

The British mathematician Clive Humby (2006) defined "Data as the new oil" and Palmer (2006) extended this concept with the following sentence: "Data is just like crude. It's valuable, but if unrefined it cannot really be used. It has to be changed into gas, plastic, chemicals, etc. to create a valuable entity that drives profitable activity; so, must data be broken down, analyzed for it to have value." The Italian cognitive scientist Scaruffi (2019) analyzed the difference between oil and data considering that "the product of oil does not generate more oil (unfortunately), whereas the product of data (self-driving cars, drones, wearables, etc.) will generate more data (where do you normally drive, how fast/well you drive, who is with you, etc.)."

But despite the huge amount of data "the paradox of the great civilization change consists in the fact that we have practically unlimited access to information and data and yet we are nearly unable to use it in any way" (Castells, 2009). Consequently, models can be considered the new gold.

The inclusion of spatial simulation techniques in recent GIS software favored the diffusion of these methods, but in several cases led to the mechanism based on which buttons have to be pressed without having geography or processes in mind.

Spatial modeling, analytical techniques, and geographic analyses are therefore required in order to analyze data and to facilitate the decision process at all levels, with a clear identification of the geographic information needed and reference scale to adopt.

But knowledge is not owned by someone, but it must be shared to the public.

4.5 Public participation and knowledge sharing

In recent times, a lot of activities have been developed with the support of mass cooperation. This tendency often adopted in public agencies and local authorities is based on an open government approach. This concept is based on a more participative method to government where citizen's ideas and activities have to be considered and collected in a sort of a continuous flow. Consequently, public involvement, collecting ideas, suggestions, or simply data/information production, is a daily activity considered as fundamental in decision-making. Obama's administration has given a great impetus to this approach, implementing such a policy and enlarging the possibility to capture public imagination by means of social networks, blogs, and all possible solutions for directly interacting with citizens (O'Reilly, 2009a).

This new approach is often called Gov. 2.0 (Jaeger, Bertot, & Shilton, 2012; O'Reilly, 2009b). Open government (Lathrop & Ruma, 2010) without a

2.0 approach is still based on a direct action. "Providers" is a sort of Right to Information where the administration tries to inform people but having interaction just with the main stakeholders. Gov. 2.0 is a more open approach, which "enables" citizens to have an important role in defining policies as well as in producing user-friendly, ubiquitous, and personalized services.

Social media and all 2.0 platforms are key elements in generating direct contact with citizens. Extensions of 2.0 philosophy have changed the relationship between citizens and administrations completely. People directly realize services that public administration is not interested to carry out, and the private sector does not consider convenient to realize.

Web 2.0 tools, such as WEBSITE, BLOG, WebGIS, and mobile applications, for smartphone and tablet represent a sort of transition from "one-way" to "two-way" information and interaction tools able to share ideas, compare opinions, and collect information (Evans-Cowley & Conroy, 2006; Murgante et al., 2019; Murgante, Tilio, Lanza, & Scorza, 2011).

In 2.0 planning approaches, citizens, probably unconsciously, face many of the typical steps adopted in the planning process, reaching the highest level of Arnstein, Kingston, and Haklay Ladders. In terms of planning theory, 2.0 planning can be seen as a renewed approach to Advocacy Planning (Davidoff, 1965), where the collector of instances is a virtual environment. The advocacy planner represents all people generally unheard in the decision process, and the needs of marginalized neighborhoods, generally absent at decision-making tables. While in advocacy organizations, a sort of hierarchy remains between the mass of people and their representatives, in 2.0 planning all people have the same position on a scale of responsibility. The development of 2.0 planning is strictly related to social media growth. Facebook, Twitter, and the other social networks were born to look for old classmates, military service friends, and university colleagues; today these are powerful media and places where it is possible to exchange ideas and opinions (Rocha, Pereira, & Murgante, 2015; Rocha, Pereira, Loiola, & Murgante, 2016; Resch, Summa, Zeile, & Strube, 2016).

The use of social networks has enabled a significant expansion of participatory basis, beyond the constraints of space and time (Salvini, 2005). Social scanning (Pang, 2010) is a fundamental instrument for collecting ideas, opinions, etc., from citizens. Social platforms can lead from a closed model of decision-making based on professionals' government and representative democracy, where participation is mainly relegated to election (Noveck, 2009) to an integration of representative democracy and collaborative approaches in which a decision maker has the possibility to directly consult citizens in order to make a particular decision. If, on one side, it is important to avoid pitfalls highlighted by Michael Bloomberg in an interview to the New York Times (Grynbaum, 2012), where Twitter has been defined as a source of everyday referendum leading activities only to short-term actions because great part of people is not interested in future programming activities; on the other side, social platforms can produce social mobilization, claims, and real changes in people quality of life (Healey, 2001).

This approach generated all over the world smart communities. Murgante and Borruso (2015) defined the three main pillars of a smart city:

1. Connections—as networks and technological infrastructures.
2. Data—open and public or public interest data to allow the development of innovative solutions and the interaction between users/citizens and the city.
3. Sensors—have to be intended not only in terms of technology, but also in terms of citizens to be able to actively participate in a bottom-up way to city activities and data production.

Such pillars need to be accompanied by an urban governance to be able to harmonize them and particularly to represent a set of minimum "driving rules," regulating a smart city in a neutral way, without entering too much into details concerning contents and applications developed by citizens, urban users, private companies, local associations, etc.

In such terms, a true smart city acts as an enabling platform for the activities that citizens are able to develop, linking those inherited from the past to those that can be realized in the future, so it is not focused just on applications but also on the possibility that citizens realize them.

But, now the problem is: how to practically deal with knowledge?

4.6 Rules for urban planning

Rules constitute a very important component of knowledge. In artificial intelligence, the representation of rules is based on several mathematical theories, such as classical logics. Moreover, according to Graham (2006) and Morgan (2008), rules should be considered as first-class citizens in computer science. In this section, first, generalities about geospatial rules are examined, and then some preliminary elements to get them machine processable are given.

4.6.1 Generalities about geospatial rules

Consider some examples of knowledge chunks in smart cities, some of them coming from physical laws, administrative regulations, or best practices (Laurini, 2020):

- When planning a metro, move underground networks.
- Each building must be connected to utility networks (water, electricity, gas, telephone, Internet, etc.).
- If a crossroad is dangerous, install traffic lights.
- In city centers, transform streets into pedestrian precincts.
- When a commercial mall is planned in the neighborhood of a city, shops located in the city center will be in jeopardy.
- If the number of car parking lots is insufficient, encourage using buses or bikes.
- At the vicinity of an airport, limit building heights.
- When a big plant is closing, unemployment will increase.

- At the vicinity of an historic building (listed monument), no modifications of building are allowed.
- When defining a new industrial area, unemployment will diminish.
- If a recreational park is inside a city, provide bike lanes coming to this park.
- In France, it is forbidden to open a new tobacconist shop within 500 m from an existing one.
- If there is one or several rivers crossing a city, design systems to mitigate floods.
- In a city with many hills, consider cable cars linking them.

Unlike business rules, which are encoded with logic, those geospatial rules need to integrate computational geometry, topology, and operation research (especially for looking for optimum).

Generally, the implementation of rules is based on two grammatical structures: IF-THEN-Fact and IF-THEN-Action (Ross, 2011). The first serves above all to involve new facts, that is, for us, new objects, attribute values, and new relationships between geographic objects. And the second is to involve new actions. But who will be in charge of such new actions? In some cases, the computer itself may run procedures; in others, particularly in regulatory contexts, a decision maker (for example, the Mayor of a municipality) must himself initiate the action. Another interpretation could be the choice of alternatives of an action, for example, when a law, in some well-defined contexts, opens many perspectives.

4.6.2 Languages for rule encoding

Several languages to model rules exist. For instance, Boley, Paschke, and Shafiq (2010) suggested several XML extensions to model rules. The simplest of these is as follows:

```
<Implies>
        <if>
                <..>
        </if>
        <then>
                <..>
        </then>
</Implies>
```

Laurini (2019) proposed a mathematical language for rule encoding, but not yet a computer language, essentially because it depends on the structuring of geographic knowledge base and system in use. This language can be simplified with the following specifications:

- Antecedents will be represented as a context with quantifiers ("∃" or "∀") followed by the symbol ":" and some Boolean expressions to model conditions.
- The symbol "⇒" when the implication is mandatory.
- Consequents (as acts or actions); if there are many, they will be parenthesized by "{" and "}," and each separated by ";".

But rules will be based on different elements which constitute a spatial knowledge base, namely, geographic objects (identifiers, toponyms, geometry, and attributes), gazetteer, ontologies, external knowledge, etc. (see Laurini, 2017 for more details). Consider a place *P*, we will denote its geometry by *geom* (*P*), and its name by *Topo* (*P*). In addition, there are conventional topological relations (contains, meet, etc.) and geometric functions (union, centroid, buffer, etc.).

4.6.3 Example: Projected buildings and urban planning rules

To illustrate, let us take an example in planning zones as depicted in Fig. 6, each with its own regulations. For instance, campsites are not allowed in Downtown. To simplify, three zones are defined, Downtown (*ZoneA*), suburban area (*ZoneB*), and rural area (*ZoneC*). However, historical monuments are located downtown, and it is forbidden to construct new buildings around them (for instance, within 200 m). In the rural area, there is an airport for which some limitations exist.

Let us begin by *ZoneB*, which is the simpler rule in which building height is limited to 15 m.

∃ *C* ∈ *City*, ∃ *ZoneB* ∈ *Terr*, ∀ *B* ∈ *Project.Building*, ∀ *P* ∈ *Parcel*, *Topo* (*C*) ≡ "Smart Town," *Topo* (*ZoneB*) ≡ "Suburban Area" ⊧ *Contains* (*Geom* (*C*), *Geom* (*ZoneB*)), ⊧ *Contains* (*Geom* (*ZoneB*), *Geom* (*P*)), ⊧ *Contains* (*Geom* (*P*), *Geom* (*B*)) : *B.Height* ≤ *15* ^ *Area* (*Union* (*Geom* (*Floors*)))/*Area* (*Geom* (*P*)) ≤ 4 ^ *Area* (*B*)/ *Area*(*Geom* (*P*)) ≤ 0.70 ⇒ ⊧ *B.ZoneB_Approved* ∎	Rule 1

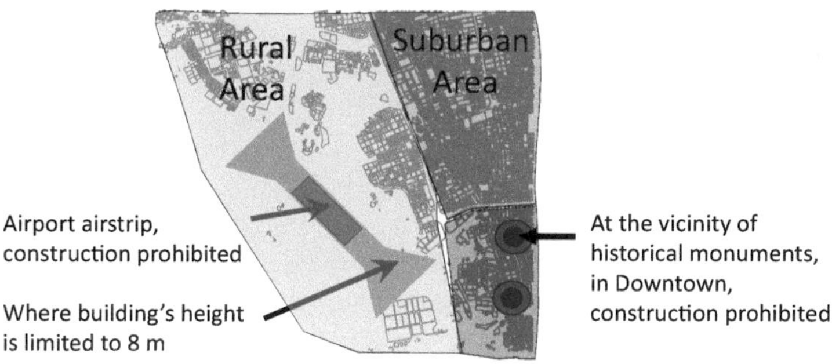

FIG. 6 Planning zones.

For Downtown (*ZoneA*), it is more complicated because of the buffer zones of 200 m around historical buildings.

$$
\begin{gathered}
\exists\, C \in City,\ \exists\, ZoneA,\ ConservA \in Terr,\ \forall\, B \in Project.Building,\\
\forall\, P \in Parcel,\ \forall\, M \in Monuments,\\
Topo\ (C) \equiv \text{"Smart Town,"}\\
\text{Topo}\ (ZoneA) \equiv \text{"Downtown,"}\\
Geom\ (ConservA) \equiv Union\ (Buffer\ (Centroid\ (Geom(M),\ 200))),\\
\models Contains\ (Geom\ (C),\ Geom\ (ZoneB)),\\
\models Contains\ (Minus\ (Geom\ (ZoneA),\ Geom\ (ConservA)),\ Geom\ (P)),\\
\models Contains\ (Geom\ (P),\ Geom\ (B))\\
:\\
B.Height \leq 12\\
\wedge\, Area\ (Union\ (Geom\ (Floors)))/Area\ (Geom\ (P)) \leq 3\\
\wedge\, Area\ (B)/\ Area(Geom\ (P)) \leq 0.80\\
\Rightarrow\\
\models B.ZoneA\_Approved \blacksquare
\end{gathered}
$$

Rule 2

For the Rural Area, taking the airport into consideration, we need to consider three areas: the area outside the airport, the airstrip in which any building is forbidden, and finally in the "bowtie" with additional limits (Rule 3). Technically speaking, we have to use an "exclusive or," noted ⊕ between those three possibilities.

$$
\begin{gathered}
\exists\, C \in City,\ \exists\, ZoneC,\ Bowtie,\ Airstrip \in Terr,\\
\forall\, B \in Project.Building,\\
\forall\, P \in Parcel,\\
Topo\ (C) \equiv \text{"Smart Town,"}\\
Topo\ (ZoneA) \equiv \text{"Rural Area,"}\\
Geom\ (Bowtie) = Polyg\ (640, 243;\ 657, 290;\ 748, 387;\ 796, 405;\ 743,\\
459;\ 729, 406;\ 636, 316;\ 580, 297),\\
Geom\ (Airstrip) = Polyg\ (670, 311;\ 724, 365;\ 707, 386;\ 650,\\
330),\\
\models Contains\ (Geom\ (C),\ Geom\ (ZoneC)),\\
\models Contains\ (Geom\ (P),\ Geom\ (B))\\
:\\
(Contains\ (Minus\ (Geom\ (ZoneC),\ Geom\ (Bowtie)),\ Geom\ (B)),\\
\wedge\, B.Height \leq 12 \wedge Area\ (Union\ (Geom\ (Floors)))/Area\ (Geom\ (P)) \leq 0.5\\
\wedge\, Area\ (B)/\ Area\ (Geom\ (P)) \leq 0.30\ )\\
\oplus\ Disjoint\ (\text{Geom}\ (Airstrip),\ Geom\ (B))\\
\oplus\ (Contains\ (Geom\ (Bowtie)),\ Geom\ (B))\\
\wedge\, B.Height \leq 8 \wedge Area\ (Union\ (Geom\ (B.Floor)))/Area\ (Geom\ (P)) \leq 0.5\\
\wedge\, Area\ (B)/\ Area\ (Geom\ (P)) \leq 0.30)\\
\Rightarrow\\
\models B.ZoneC\_Approved \blacksquare
\end{gathered}
$$

Rule 3

5. Conclusions

An article published in The Wall Street Journal explains how US major metropolitan areas produce a higher GDP than the economies of entire nations. Urbanization is also different in terms of city size classes in the two areas. In

Europe, 67% of urban inhabitants live in medium-size urban centers, less than 500,000 inhabitants, while just 9.6% are located in cities having more than 5 million inhabitants. In the United States, one out of five urban inhabitants lives in major cities having more than 5 million people. Consequently, it is very easy to understand that, despite common opinions against the quality of life in big cities, in most cases living in large cities becomes a necessity. Glaeser (2011) defines the city as the greatest invention of mankind. Using the advantages of the agglomeration principle, a city emphasizes the strengths of a society. Despite the evolution of modern and contemporary cities having led to disadvantages resulting from congestion, urban poverty, and security, living today in an urban context, even one that is not of high quality, involves more benefits than living in remote areas. Consequently, cities play a central role for humanity, offering the opportunity to learn from each other, face to face. Despite economic contexts and production patterns having been radically changed, a city always represents the most vital element of the economy of a nation. Generally, in every developed country, cities are the economic heart and the most densely populated places, very attractive for people who want to exchange or build knowledge because cities have the capability to provide something for everybody, only because, and only when, they are created by everybody (Jacobs, 1961). Batty (2013) defines cities as kaleidoscopes of plurality, a multiplicity of ideas, perceptions, theories, and models. Consequently, cities can be considered as an enabling environment where it is possible to create, build, and exchange knowledge.

Most recent events seem to put in crisis that point of view, even if it is day by day, it is diffused and appreciated as we consider cities also a concentration of problems, and a hypothesis to check whether a good relationship is possible between city and nature or scattered settlements. As complexity that this question poses, is truly relevant, any effort has to be produced by scientific research; maybe an effort is needed to deal with a renovated approach to a large-scale planning, i.e., a renovated approach to strategic planning where actions are strictly depending on objectives, and spatial transformation to economy development. In particular Archibugi (2008) declares that the goal is a well-founded activity based on an accountable approach that respects carefully the legitimacy of the request to use resources in a rational way, i.e., the best compromise among equity, efficiency, and natural resources conservation. Finally, the path to the city of knowledge will be a sort of cross-fertilization between economy, education, and urban planning based on public participation.

References

Afgan, N. H., & Carvalho, M. G. (2010). The knowledge society: A sustainability paradigm. *Cadmus*, *1*(1). http://www.cadmusjournal.org/node/14.

Angelidou, M., Gountaras, N., & Tarani, P. (2012). Engaging digital services for the creation of urban knowledge ecosystems: The case of Thermi, Greece. *International Journal of Knowledge-Based Development*, *3*(4), 331–350.

Angelidou, M., & Mora, L. (2019). Exploring the relationship between smart cities and spatial planning: Star cases and typologies. In *Smart cities in the post-algorithmic era* (pp. 217–234). Edward Elgar. https://doi.org/10.4337/9781789907056.00019.

Archibugi, F. (2008). Planning theory: From the political debate to the methodological reconstruction. In Springer (Ed.), *Planning theory*. Mailand: Springer-Verlag. https://doi.org/10.1007/978-88-470-0696-6_1.

Batty, M. (2013). *The new science of cities*. Cambridge, MA: The MIT Press.

Bindé, J., & Matsuura, K. (2005). *Towards knowledge societies*. UNESCO Publishing, ISBN:9789231040009.

Boley, H., Paschke, A., & Shafiq, O. (2010). RuleML 1.0: The overarching specification of web rules. In B. H. Springer (Ed.), *Vol. 6403. Semantic web rules. RuleML 2010* (pp. 162–178). Berlin, Heidelberg: Springer. https://doi.org/10.1007/978-3-642-16289-3_15.

Castelfranchi, C. (2007). Six critical remarks on science and the construction of the knowledge society. *Journal of Science Communication*, *6*(4), 1–3. http://jcom.sissa.it/.

Castells, M. (2009). The rise of the network society. In *The rise of the network society, Volume I: Second edition with a new preface* Wiley. https://doi.org/10.1002/9781444319514.

Chadwick, G. F. (1978). *A systems view of planning: Towards a theory of the urban and regional planning process*. Oxford, UK: Pergamon Press.

Davidoff, P. (1965). Advocacy and pluralism in planning. *Journal of the American Planning Association*, *31*(4), 331–338. https://doi.org/10.1080/01944366508978187.

Diderot, D. (1751). https://encyclopedie.uchicago.edu/content/syst%C3%A8me-figur%C3%A9-des-connaissances-humaines-0.

Edwards, P. N., Jackson, S. J., Chalmers, M., Bowker, G., Borgman, C. L., Ribes, D., et al. (2012). *Knowledge infrastructures: Intellectual frameworks and research challenges*. Report of a workshop sponsored by the National Science Foundation and the Sloan Foundation University of Michigan School of Information, 25-28 May 2012 http://pne.people.si.umich.edu/PDF/Edwards_etal_2013_Knowledge_Infrastructures.pdf.

Evans-Cowley, J., & Conroy, M. M. (2006). The growth of e-government in municipal planning. *Journal of Urban Technology*, *13*(1), 81–107. https://doi.org/10.1080/10630730600752892.

Faludi, A. (1987). *A decision-centered view of environmental planning*. Oxford, UK: Pergamon Press.

Fleurbaey, M. (2018). *Priority to the furthest behind*. www.un.org/development/desa/dpad/cdp-back.

French Environment Charter. (2004). https://www.legifrance.gouv.fr/contenu/menu/droit-national-en-vigueur/constitution/charte-de-l-environnement.

Geddes, P. (1915). *Cities in evolution; an introduction to the town planning movement and to the study of civics* (New edition 1968). H. Fertig.

Glaeser, E. (2011). *Triumph of the city: How our greatest invention makes us richer, smarter, greener, healthier, and happier*. Penguin Random House. https://www.penguinrandomhouse.com/books/303439/triumph-of-the-city-by-edward-glaeser/.

Graham, I. (2006). *Business rules management and service oriented architecture: A pattern language*. Hoboken, NJ: Wiley.

Grynbaum, M. M. (2012). Mayor warns of the pitfalls in social media. *The New York Times*, (March 21). https://www.nytimes.com/2012/03/22/nyregion/bloomberg-says-social-media-can-hurt-governing.html?auth=login-facebook.

Healey, P. (2001). Editorial. *Planning Theory and Practice*, *2*(3), 377–380. https://doi.org/10.1080/14649350120096802.

Jacobs, J. (1961). *The death and life of great American cities*. New York, NY: Random House.

Jaeger, P. T., Bertot, J. C., & Shilton, K. (2012). Information policy and social media: Framing government—Citizen web 2.0 interactions. In *1. Public administration and information technology* (pp. 11–25). Springer. https://doi.org/10.1007/978-1-4614-1448-3_2.

Las Casas, G., Scorza, F., & Murgante, B. (2019). Razionalità a-priori: una proposta verso una pianificazione antifragile. *Italian Journal of Regional Science, 18*(2), 329–338. https://doi.org/10.14650/93656.

Lathrop, D., & Ruma, L. (2010). *Open government*. O'Reilly Media. https://www.oreilly.com/library/view/open-government/9781449381936/.

Laurini, R. (2017). Geographic knowledge infrastructure: Applications to territorial intelligence and smart cities. In *Geographic knowledge infrastructure: Applications to territorial intelligence and smart cities*. Amsterdam, The Netherlands: Elsevier.

Laurini, R. (2019). A mathematical language for the modeling of geospatial static rules. *Journal of Visual Language and Computing, 1*(1), 1–13. https://doi.org/10.18293/JVLC2019-N1.

Laurini, R. (2020). A primer of knowledge management for smart city governance. *Land Use Policy*, 0264-8377. *2020*, 104832. https://doi.org/10.1016/j.landusepol.2020.104832.

Lytras, M. D., & Sicilia, M. A. (2005). The knowledge society: A manifesto for knowledge and learning. *International Journal of Knowledge and Learning, 1*(1–2), 1–11. https://doi.org/10.1504/ijkl.2005.006259.

McLoughlin, J. B. (1969). *Urban and regional planning: A systems approach*. London, UK: Faber & Faber.

Morgan, T. (2008). *Business rules and information systems: Aligning IT with business goals*. Boson, MA: Addison-Wesley.

Murgante, B., & Borruso, G. (2013). *Cities and smartness: A critical analysis of opportunities and risks* (pp. 630–642). Berlin, Heidelberg: Springer.

Murgante, B., & Borruso, G. (2014). *Smart city or Smurfs city* (pp. 738–749). Springer International Publishing.

Murgante, B., & Borruso, G. (2015). Smart cities in a smart world. In *Future city architecture for optimal living* (pp. 13–35). Heidelberg, Germany: Springer International Publishing.

Murgante, B., Botonico, G., Graziadei, A., Sassano, G., Amato, F., & Scorza, F. (2019). Innovation, technologies, participation: New paradigms towards a 2.0 citizenship. *International Journal of Electronic Governance, 11*(1), 62–88. https://doi.org/10.1504/IJEG.2019.098814.

Murgante, B., Tilio, L., Lanza, V., & Scorza, F. (2011). Using participative GIS and e-tools for involving citizens of Marmo Platano-Melandro area in European programming activities. *Journal of Balkan and Near Eastern Studies, 13*(1), 97–115. https://doi.org/10.1080/19448953.2011.550809.

Noveck, B. S. (2009). *Wiki government: How technology can make government better, democracy stronger, and citizens more powerful*. Washington, DC: Brookings Institution Press.

O'Reilly, T. (2009, April 27). *Obama's New Tech Guru*. Forbes. https://www.forbes.com/2009/04/26/aneesh-chopra-innovation-technology-breakthroughs-chopra.html#47bf3b47251f.

O'Reilly, T. (2009, August 10). *Gov 2.0: The promise of innovation*. Forbes. https://www.forbes.com/2009/08/10/government-internet-software-technology-breakthroughs-oreilly.html#9fb7efc3b7b4.

Palmer, M. (2006). *Data is the new oil*. https://ana.blogs.com/maestros/2006/11/data_is_the_new.html.

Pang, A. S. K. (2010). Social scanning: Improving futures through web 2.0; or, finally a use for twitter. *Futures, 42*(10), 1222–1230. https://doi.org/10.1016/j.futures.2010.09.003.

Paul, S. (2014). *Science and technology capacity and the knowledge society*. https://fr.slideshare.net/SD_Paul/science-and-technology-capacity-and-the-knowledge-society.

Resch, B., Summa, A., Zeile, P., & Strube, M. (2016). Citizen-centric urban planning through extracting emotion information from twitter in an interdisciplinary space-time-linguistics algorithm. *Urban Planning*, *1*(2), 114–127. https://doi.org/10.17645/up.v1i2.617.

Rocha, M. C. F., Pereira, G. C., Loiola, E., & Murgante, B. (2016). Conversation about the city: Urban commons and connected citizenship. *Lecture notes in computer science (including subseries lecture notes in artificial intelligence and lecture notes in bioinformatics): Vol. 9790* (pp. 608–623).

Rocha, M. C. F., Pereira, G. C., & Murgante, B. (2015). *City visions: Concepts, conflicts and participation analysed from digital network interactions* (pp. 714–730). Cham: Springer.

Ross, R. G. (2011). More on the if-then format for expressing business rules: Questions and answers. *Business Rules Journal*, *12*(4). http://www.brcommunity.com/articles.php?id=b588.

Salvini, A. (2005). *L'analisi delle reti sociali. Risorse e meccanismi*. Pisa, Italy: Pisa University Press.

Scaruffi, P. (2019). *A history of silicon valley—Vol 1: The 20th century*. Independently Published. https://www.amazon.com/dp/1686595050.

Shekhar, S., & Vold, P. (2020). Spatial computing. *MIT Press (Essential knowledge series)*.

United Nations. (2015a). *Sustainable development goals (SDGs)*. https://www.un.org/ga/search/view_doc.asp?symbol=A/RES/70/1&Lang=E.

United Nations. (2015b). *Transforming our world: The 2030 agenda for sustainable development*. https://sustainabledevelopment.un.org/content/documents/21252030 Agenda for Sustainable Development web.pdf.

United Nations Development Programme. (2018). *What does it mean to leave no one behind?*. https://www.undp.org/content/undp/en/home/librarypage/poverty-reduction/what-does-it-mean-to-leave-no-one-behind-.html.

Valet, V. (2020). *Meet America's best management consulting firms 2020. Forbes*. https://www.forbes.com/sites/vickyvalet/2020/03/17/meet-americas-best-management-consulting-firms-2020/#2653c70956ab.

Yigitcanlar, T. (2015). Knowledge based urban development. In M. Khosrow-Pour (Ed.), *Encyclopedia of information science and technology* (3rd ed., pp. 7475–7485). Hershey, Pennsylvania, USA: IGI Global. . https://www.igi-global.com/chapter/knowledge-based-urban-development/112448.

Further reading

Arnstein, S. R. (1969). A ladder of citizen participation. *Journal of the American Planning Association*, *35*(4), 216–224. https://doi.org/10.1080/01944366908977225.

Castells, M. (2000). *The rise of the network society*. Blackwell Publishers.

Haklay, M. (2013). Citizen science and volunteered geographic information: Overview and typology of participation. In *Vol. 9789400745872. Crowdsourcing geographic knowledge: Volunteered geographic information (VGI) in theory and practice* (pp. 105–122). Netherlands: Springer. https://doi.org/10.1007/978-94-007-4587-2_7.

Kingston, R. (2002). The role of e-government and public participation in the planning process. In *XVI AESOP Congress, Volos, Greece, July 10th–14th 2002*. http://www.ccg.leeds.ac.uk/groups/democracy/presentations/AESOP_kingston.pdf.

Murgante, B. (2013). Wiki-planning: The experience of Basento Park in Potenza (Italy). In G. Borruso, S. Bertazzon, A. Favretto, B. Murgante, & C. Torre (Eds.), *Geographic information analysis for sustainable development and economic planning: New technologies* (pp. 345–359). IGI Global. https://doi.org/10.4018/978-1-4666-1924-1.ch023.

Chapter 15

Foresight: A visionary step for becoming a smart city

Andrei Ştefan Neştian[a] and Ana Iolanda Vodă[b]

[a]*Faculty of Economics and Business Administration, Alexandru Ioan Cuza University, Iaşi, Romania,* [b]*Institute for Interdisciplinary Research, Social Science and Humanities Research Department, Alexandru Ioan Cuza University, Iaşi, Romania*

1. Introduction

The 2030 Agenda includes 17 Sustainable Development Goals (SDGs) and 169 associated targets aimed to provide support and capacity building for all countries. Heads of State and Government and High Representatives of 193 countries defined, in September 2015, the SDGs in order to improve health and education, reduce inequality, and spur economic growth for current and future generations (UN, 2015). This supremely ambitious and transformational vision is set to be implemented in an increasingly interdependent and multifaceted world. In this context, foresight exercises are an essential tool for identifying the risks and opportunities in the future policy implementation environment. Foresight provides future alternative scenarios for short-, medium-, and long-term paths; it gives new insights, reduces complexity and uncertainty by drawing upon and creating knowledge networks. Therefore, it is imperative to identify various viewpoints of society and technologies increasing awareness and understanding of potential and plausible paths of action and their consequences.

The SDG 11, within the Sustainable Development Agenda, is the first standalone global development goal on urban development that is focused on transforming cities and human settlements toward more inclusive, safe, resilient, and sustainable societies. The SDG 11 includes seven direct targets that cover aspects of more adequate access to safe and affordable housing and basic services (SDG 11.1), access to safer, more affordable, and sustainable transport system (SDG 11.2), and more inclusive and sustainable planning (SDG 11.3). It aims to protect the natural and cultural heritage (SDG 11.4) and the poor and vulnerable (SDG 11.5). Also, it focuses on the reduction of urban impact on the environment (SDG 11.6) and provision of access to safe, inclusive, and accessible green and public spaces for disadvantaged groups (SDG 11.7) (UN, 2015; UNESC, 2015). Making cities more sustainable means creating career

Smart Cities and the UN SDGs. https://doi.org/10.1016/B978-0-323-85151-0.00015-4

and business opportunities, providing safe and affordable housing, and building resilient societies and economies. Examples may include investment in public transport and creation of green public spaces, better urban planning and management in participatory and inclusive ways, and development of resilient approaches to pursue the preferred future (UN, 2015). To achieve the SDGs 11 target goals, we need to understand better why and how we imagine the future. It is believed that foresight brings value to policy formulation and implementation by linking knowledge about the future to the policy and strategy implementation cycle (see Table 1).

By applying foresight exercises to classic planning processes, policymakers benefit from the opportunity to address both responsiveness to change and citizen-centered service design. Also, it helps policies to increase their resilience to different future environments. As a trend-spotting tool, foresight is useful for developing possible futures which enhance classical planning and strengthen the anticipatory and adaptive government. Foresight is most often concerned with using future intelligence to plot short-, medium-, or long-range futures in uncertain and unpredictable environments to get a better understanding of the nature and pace of change and to identify potential opportunities and likely future developments. Foresight exercises provide a more effective and solid platform for broader stakeholders' inclusion and engagement.

The chapter is organized as follows: the next sections are focused on developing a conceptual framework for foresight exercises, then follows an

TABLE 1 Classic policy cycle versus foresight policy and strategy implementation cycle.

Classic policy cycle	Policy cycle with foresight
Modify objectives	Modify objectives
Formulate policy	Formulate policy
Implement policy	Implement policy
Monitor events	Monitor events
	Anticipate future events
	• Gather intelligence • Explore the dynamics of change • Describe the possible future • Develop responses
Evaluate impact	Evaluate impact

From Gov, UK (2017). The futures toolkit: Tools for futures thinking and foresight across UK government. Government Office for Science.

assessment of the importance of vision building for building smart cities and its resulted benefits. The last section presents the study conclusions.

2. Foresight exercises: A conceptual framework

Since the launch of British foresight programs (Martin, 2010), several definitions have been suggested in order to ascertain the concept of "foresight." According to the EU Commission report, foresight is a "systematic, participatory, future-intelligence-gathering and medium to-long-term vision-building process aimed at enabling present-day decisions and mobilizing joint actions" (FOREN Foresight for Regional Development Network, 2001). Visions are seen as orienting strategic operational urban planning, as well as monitoring and adapting existing plans (John, Keeler, Wiek, & Lang, 2015). Some authors explain the foresight concept in terms of the development of used methods (Reger, 2001), while others discuss it from the perspective of process aims (Martin, 1995).

Although numerous definitions of foresight have been developed over time, some common aspects should be underlined.

Foresight is seen as a process, not just as a set of techniques (Marinova, Ul-Haq, Portaleoni, & Marinov, 2013; Zeng, Koller, & Jahn, 2019). The foresight process involves gathering, understating, and assessment of information about future events and also ensures feedback to and from the procedures of relevant actors. From the approach perspective, it includes a combination of normative, explorative, or both methods, in terms of their approach. *Normative foresight assessments* look into the goals to be pursued, and which ones are wanted for the future, as well as into the proposed plan and how it can be achieved. *Explorative foresight approach* develops and explores possible futures through building scenarios, visions, and roadmaps using today's knowledge as a point of departure (Wippel, 2014). The aim of this process is to "identify the areas of strategic research and the emerging of generic technologies likely to yield the greatest economic and social benefits" (Martin, 1995). Foresight consists of multiple transformative scenarios of alternative development paths and of possible implications lying ahead.

Foresight includes engaging key stakeholders, policy-makers, or other actors so that joint vision and implementation of proper actions is facilitated. By increasing awareness of future and plausible paths, foresights provide the practical tools that can help mobilize policy-makers or other actors in line with their shared prospects and visions. According to Van der Meulen, De Wilt, and Rutten (2003), foresight can be seen as a process of "managing uncertainty through intensive interaction between stakeholders" (Van der Meulen et al., 2003). In other words, it can be viewed as a process of exchange, learning, and engagement among stakeholders and other public entities in order to broaden the horizon of participants and develop their shared perceptions of challenges and options in a world dominated by uncertainty and limited information.

By applying foresight exercises, we can enhance the decision-making capabilities of participants and reduce process complexity.

The foresight starting point assumes the existence of many possible futures. Foresight uses a mixed collection of methods designed to provide insights into the dynamics of change by looking at various possible and plausible trends and their implications for the future (Coates, 1985).

In essence, the process may be used to detect and deal with unexpected changes in a proactive manner by "systematically attempting to look into the longer-term future of science, technology, the economy and society" (Martin, 1995), as it is not focused on a single probabilistic future but on several possible scenarios which are likely to yield increased outcomes. However, the vast majority of city-level foresights look at medium-term perspective (5–20 years) and just a few support the exploration of a longer-term horizon (see Fig. 1).

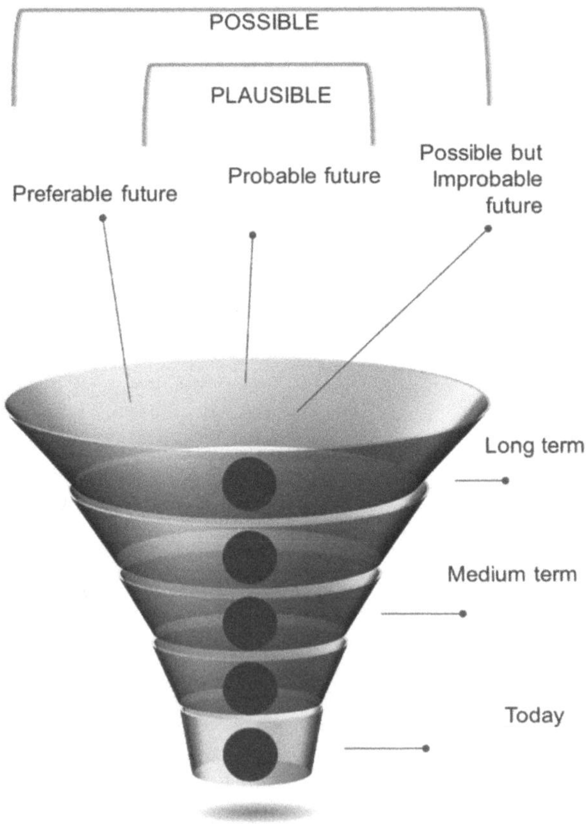

FIG. 1 City foresight diagram and different types of futures. *(From GOS (Government Office for Science) (2017). Future of cities: foresight for cities.)*

3. Sustainable future for cities: The importance of foresight exercises for smart cities

Smart cities are created in response to complex and rapid developments in urban population.

One of the most important stages in smart city development is defining the vision of a stakeholder community based on the future shape of the city. Predicting the far future needs of cities must start with a deep understanding of the existing city context and its historical development (Rogers, 2017, p. 144) and recognition of the fact that all cities are unique. This uniqueness generates the need for a customized approach to conceiving this prediction, based on the unique knowledge of those aware of the real meaning of the past and the present of a specific city. Hence, at this stage, many urban communities' resort to foresight exercises, which enables them to highlight and aggregate the needs of multiple stakeholders by suggesting innovative solutions and defining a horizon needed for developing realistic initiatives transforming the city. City foresight is "the science of thinking about the future of cities. It draws on diverse methods to give decision-makers comprehensive evidence about anticipated and possible future change" (GOS, 2017).

City policy-makers are expected to be the most interested party in developing foresight programs. It is due to the fact that city foresight presents all alternative paths and equips policy-makers with practical tools for enhancing their decision-making capabilities by using different scenario design exercises and exploratory hypotheses. City foresight offers to decision-makers extensive documentations on possible future changes in an increasingly complex and interconnected environment (GOS, 2017, p. 4).

As the socioeconomic environment becomes more complex and less predictable, a foresight approach to urban planning and technologies provides an opportunity for cities to reposition themselves for new future roles and to cope with anticipated future environmental and socioeconomic change (Dixon, Montgomery, Horton-Baker, & Farrelly, 2018). So, due to its importance for strategic planning, city foresight is useful for its wider purpose in building "strategic urban co-intelligence." Yet, a deficiency of visioning studies and foresight exercises is a detachment or mismatch between the process output and its applications in policy or enterprise (Ravetz & Miles, 2016).

Fernández Güell and López (2016) consider that one major challenge in city foresight is the complexity of city conceptualization which is achieved during the foresight exercise. They put forward three parameters for assessing this complexity: functional, territorial, and contextual. Functional complexity is proportional to the number and diversity of city operational processes, involved stakeholders, and relationships that take place within the urban systems. The higher the functional complexity, the greater is the need for splitting the foresight into several axes or topics, and the need to introduce smart city features. The territorial complexity deals with the geographical relationships which need to be considered as relevant for understanding how the city operates. Assessment of territorial complexity might be done using three different scales: metropolitan, regional, and global (Fernández Güell & López, 2016).

Several perspectives have been developed to describe the steps, phases, or stages in a city foresight initiative. According to Fernández Güell and López (2016), the development of future paths can be implemented through five sequential steps (see Fig. 2).

This methodological framework recognizes the complexities that emerge with the development of smart city initiatives. Its application is divided into two approaches, systemic and foresight, which incorporates five subsequent and necessary steps.

In the first step, smart city functionalities are drawn by using a simplified conceptual map. The schematic representation assesses a synthesized set of different interrelated subsystems which may affect the city functioning balance more likely simultaneously than in a linear sequence. In the second step, the system validity is tested in a specific urban situation, then, the local stakeholders examine its legibility. In step 3, a foresight exercise is applied, which offers plausible and probable paths for future events in cities and presents the necessary tools that stakeholders and policy-makers can use to support the development of a policy that is resilient to a range of attainable results. The urban futures diminish uncertainty, provide guidance, and manage sustainable transitions toward a desirable path (McPhearson, Iwaniec, & Bai, 2016). In step 4, possible consequences are drawn from the future vision into the form and operations of the urban functional system. Step 5 assesses the importance of stakeholders in the exchange process and knowledge dissemination. Through mutual

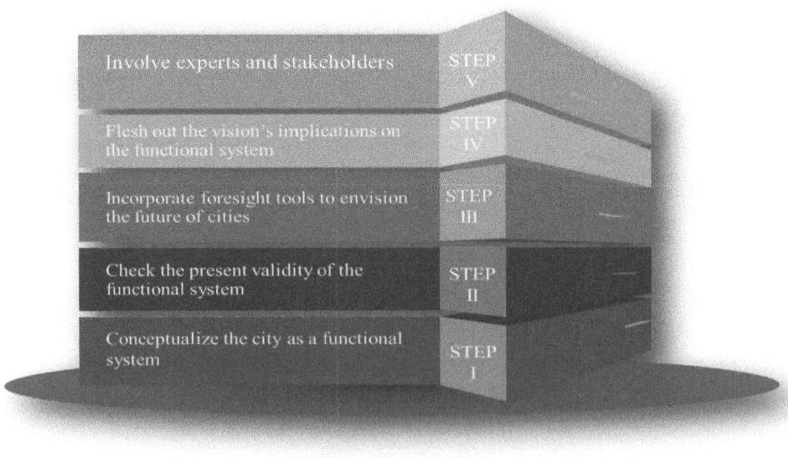

FIG. 2 Methodological framework assessment. *(From Fernández-Güell, J.M., Collado-Lara, M., Guzmán-Araña, S., & Fernández-Añez, V. (2016). Incorporating a systemic and foresight approach into smart city initiatives: The case of Spanish cities. Journal of Urban Technology, 23(3), 43–67.)*

TABLE 2 Stages in implementing foresight.

Main stages in implementing foresight exercises	Recommendations
1. Exploring aspirations and visions	The first stage is to invite stakeholders to launch themselves into future-oriented mind-sets, searching for the most distant and least familiar-future they can hope for. Exploring visions and aspirational values is one of the most accessible starting points
2. Understanding histories and now	The second stage is to raise awareness about the historical influences that have shaped a city's definitive factors. It is a way for highlighting trends and sequences of events that characterize the current state and context of a city
3. Exploring future trends	The third stage is exploring possible changes over a period of up to 15–20 years into the future
4. Considering future options	The fourth stage involves creating future policy or development options to cover the gap between now and the distant future, such as programs, investments, interventions, and actions
5. Testing future options	This stage analyses and clarifies decisions considering the uncertainties and risks associated with the different options and future outcomes of those
6. Connecting with roadmaps	A final stage is to envision the roadmap that will link the current state of a city with its future aspirations, highlighting the preferred policy options. Alternative "roadmaps" or scenarios are usually conceived

From Gov, UK (2017). The futures toolkit: Tools for futures thinking and foresight across UK government. Government Office for Science.

learning and knowledge exchange, stakeholders deepen their understating of the driving forces that affect future developments, identify gaps, and build new strategies that are more flexible to changing conditions in cities. The resulted outcomes give a more robust perspective to the interrelated complexities of cities (Fernández Güell & López, 2016).

In implementing foresight exercises in cities we should take into consideration the following steps: exploring aspirations and visions; understanding histories and the present moment; exploring future trends; considering future options; and testing future options and connecting them to roadmaps (Gov. UK, 2017). The main steps are discussed in detail in Table 2.

3.1 Themes relevant for smart city foresight exercises

High level of abstraction involved in modeling the complex perspectives to a smart city leads to a necessary breakdown of the problems into several distinct categories or themes to be treated during the foresight. The same complexity and the different visions of practitioners and theorists in the field make it difficult to reach a common perspective on the list of relevant themes for smart city foresight exercises. In the following pages, we will preset four different approaches.

Starting from a framework set forth in a report of the UK Government Office for Science (GOS, 2017), we have added arguments for a list of six main themes considered as relevant for city foresight exercises: (1) living in cities; (2) urban economies; (3) urban metabolism; (4) urban form; (5) urban infrastructure; and (6) urban governance.

1. Living in cities
   Cities are nowadays the nexus of retail, consumption, arts, and public services incorporating a wide variety of factors and areas which interrelate with each other. Over the last 50 years, cities have undergone a series of major changes due to economic restructuring, job mobility, living conditions, better employment, education opportunities, and higher income, among others. However, a major concern has been arising in most cities related to the substantial increase of 75+ aged population and shrinkage of younger broad age groups. In order to diminish the potential negative effects, governments must design future policies that are directed toward assuring the same level of well-being to all citizens, regardless of their age. Future developments in the policy area must improve the overall quality of citizens' life and include more active participation of the elderly. Polices must be designed in such a manner that unrestricted access to their available tools is provided, also being complemented by supporting initiatives encouraging longer and healthier life for the citizens (Visvizi & Lytras, 2019).
2. Urban economies
   Smart cities are designed by using smart technologies that enable us to transform and interconnect the urban ecosystem and create new business opportunities, which in turn, increase the overall economic performance of the city. Although smart cities are constantly trying to improve their growth performance and productivity, evidence shows that there are major discrepancies across cities worldwide (Gov. UK, 2017; Pérez-del Hoyo, Andújar-Montoya, Mora, & Gilart-Iglesias, 2019). This is why understating the major factors that are responsible for this outcome is imperative. The literature suggests that four major factors contribute to the different levels of economic performance due to their influence on urban economic challenges and developed policies (Gov. UK, 2017): city size, economic specialization, competitiveness, and human capital. These factors are explained in detail below:
   (a) *City size, or the accumulation of productive activities*. Evidence shows that across the European Union (EU) there is an inclination toward concentrating innovative and productive activities within the core of the city. For instance, countries gather productive activities mainly within their

core regions, while regions concentrate the cores in urban areas (Farole, Rodríguez-Pose, & Storper, 2009; Voda & Radu, 2018). However, the city size accumulation effects are not so clearly evaluated. In the literature, different authors acknowledge the importance of firm agglomeration in cities for growth (positive externalities), but this perspective is only valid in the short run. Still, it is not clear in short run whether activity concentration is beneficial or not for the city (Gov. UK, 2017).

(b) *Economic specialization*. Although economic specialization may be essential for maximizing growth, there are, of course, certain limitations on the degree to which this phenomenon contributes to the increase of city overall performance. For instance, on the long-run, a city specialized in heavy capital-intensive sectors may be less adaptable than a city specialized in knowledge-intensive activities. Therefore, flexibility appears well-correlated with firm performance as it represents the firm capacity to respond effectively to the changing outward environment (Phillips, Chang, & Su, 2019). The existence of diversity in cities allows access to a wide pool of skills and talents and to a large variety of market offers, and to relatedness of their knowledge and technologies (Gov. UK, 2017). Although specialized and diversified cities coexist, the latter is difficult to estimate, as empirically, the measurement is linked to particular classification schemes and taxonomies (e.g., number of professions in a city) that are influenced by the interpretative process and subjective views of a researcher (Duranton & Puga, 2001; Gov. UK, 2017).

(c) *Competitiveness of a city*. This factor is particularly important as it determines how adaptable a city really is and shows its capacity to develop new industries, skills, and technologies and to reconfigure its economic structure (Cetindamar, Lammers, & Sick, 2020).

(d) *Human capital*. An important quality of a smart city is to create better social and economic conditions in order to compete with other cities worldwide in attracting the best people and companies (Matos, Vairinhos, Dameri, & Durst, 2017). Therefore, education is critical in making a city attractive, especially tertiary education contributing to regional growth by increasing skills, supporting innovation, and attracting investment and talent (Matos et al., 2017; Nam & Pardo, 2011). However, a major concern for cities nowadays is not only how high-skilled labor force can be attracted but also how an appropriate mix of skills could be retrained beyond the provision of employment opportunities (Gov. UK, 2017).

3. Urban metabolism

The concept refers to the inflows of materials and energy resources, the outflows of wastes and emissions, and the retention of materials as stock in the building environment and infrastructure (Gov. UK, 2017). A green city ecosystems and green-oriented management are vital to urban resilience and livability (Duinker et al., 2015). Some city ecosystem services can include heat mitigation (Norton et al., 2015), energy saving (Komninos, Bratsas,

Kakderi, & Tsarchopoulos, 2019), and air pollution reduction (Hoq, Alam, & Amin, 2019).

4. Urban form
   It describes the city layout (streets, buildings, and open spaces) that shapes both human behavior and social outcomes. For instance, the proximity to the city center facilitates connections between citizens, which helps maintain social and economic interactions (Gov. UK, 2017).
5. Urban infrastructure
   The concept refers to buildings, agglomeration, and sprawl management with the ICT (Anthopoulos, 2015) and supports the functioning of society. Urban infrastructure consists of two parts, complemented by each other: the *hard infrastructure* providing energy and water resources for a society to function, and the *soft infrastructure* supporting people, information, and goods to move more efficiently and safely (Gov. UK, 2017).
6. Urban governance
   It regards the process through which local government and stakeholders make decisions on how to plan, finance, and manage the urban climate. Urban governance is important for the following reasons: it has an impact on the efficiency of delivered local public services; its own activities ensure cost-effective use of available resources; it shapes both physical and social character of the urban area; it may provide support and engage local residents in decision-making processes; the functions of government can change the target of accountability; and the way it responds to feedback from local citizens.

In another perspective, Nenciu, Van Der Duin, Janssen, Snijders, and Hulstijn (2017) identify four axes regrouping the uncertainties in developing city foresight: city governance, data openness, acceptance of technology, and city drivers (see Fig. 3).

- *Governance of the city: bottom-up versus top-down approach.* A bottom-up approach indicates a decentralized system with a collaborative environment, while a top-down approach refers to a centralized system with a central entity.
- *Data availability: open versus close data.* The possibility that every citizen can access and use data enables a transparent system, where costs and benefits are visible to the general public. As such, the system allows individuals to make conscious decisions related to their consumer habits. In a closed system, large companies collect and own all data generated by citizens and urban services. Data analysis and other derivative services are traded on the data market (Nenciu et al., 2017)
- *Acceptance of technology: low versus high technology acceptance.* The quality of the available ICT technology and infrastructure is one of the main features of a smart city as it provides a solid foundation for information sharing, collaboration, and interoperability that allows urban development

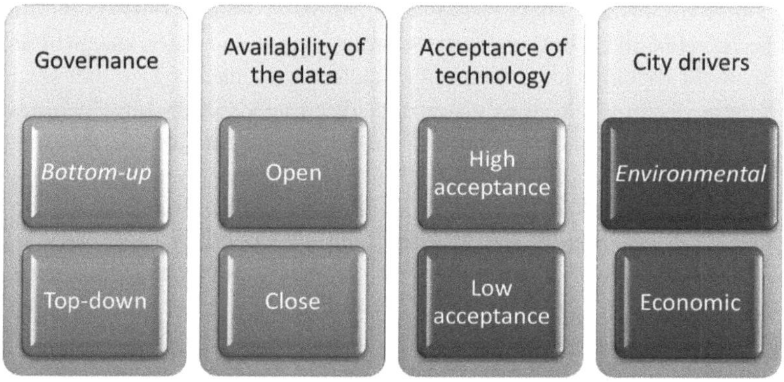

FIG. 3 Regrouping the uncertainties in developing city foresight. *(From Nenciu, G., Van Der Duin, P., Janssen, M., Snijders, D., & Hulstijn, J. (2017). Future scenarios of data—Driven smart cities. In: Proceedings of the 18th annual international conference on digital government research (pp. 471–476).)*

and welfare (Nam & Pardo, 2011; Schaffers et al., 2011). However, a future scenario where smart people are reluctant to technological advancements seems unlikely to take place (Nenciu et al., 2017).

- *Main drivers of the city—environmental versus economic drivers.* Environmental drivers relate to climate change, pollution, etc., while economic drivers can be associated with financial gains, etc.

In a third perspective, Urban Foresight Report (2016) provides a list of objectives for the smart cities within the alliance that allows us to understand their perspective on the axes of analysis recommended in the city foresight: improving lives, collaboration and engagement, open data and transparency, technology and innovation, and environmental sustainability.

The first objective assesses smart and healthy living. This covers a broad range of activities, including topics about promoting healthy living, independence, confidence, and self-esteem; tackling inequality; enhancing participation and access to public services; giving children the best start in life; and creating opportunities for citizens to succeed.

The second objective is related to the collaboration and engagement platforms of cities. This is a strategic line related to sharing resources, expertise, and rewards, as well collectively engaging citizens, decision-makers, and solution-providers that will support the successful implementation, scale-up, and commercialization of smart city projects.

The third objective relates to city's engagement in the data community and transparency. This component refers to the better provision of data and information to businesses, citizens, and public bodies. Open data is recognized as an engine for innovation, growth, and transparent governance, developing innovative smart city solutions.

Another important aspect relates to technology and innovation within cities. The focus is on embracing new technology and innovation that will enable cities to work together for delivering infrastructure and services that are more intelligently responsive to the current and future needs of citizens by attracting investment and creating opportunities to test, develop, and commercialize smart mobility solutions.

The last objective assesses environmental sustainability through the circular economy. The focus is on the use of a "circular economy" approach to efficiently manage the flow of resources within and between cities, considering that cities have a critical role in the transition to low carbon and resource-efficient economy (Urban Foresight, 2016).

Among the 17 SDGs, Goal 11 is to "Make cities and human settlements inclusive, safe, resilient and sustainable" (UN, 2015). A smart, inclusive, safe, and resilient city is an innovative city, which uses information and communication technology to enhance efficiency and connectivity, and improve the overall quality of life while taking into consideration present and future economic, social, environmental needs of society and individuals (UNESC, 2015). A Smart Sustainable City (SSC) incorporates both technological change and sustainable development goals as to enhance the longevity of economic and social progress. The SDGs 11th goal includes seven direct targets that we also view as defining relevant themes for smart city foresight exercises:

- *Housing*. Improved access to safe and affordable housing. This target is essential in preventing disasters among poor urban residents (11.1).
- *Sustainable transport system*. This target assures that city residents have access to a safer and more accessible transportation system (11.2).
- *Sustainable settlement*. This target aims at enhancing inclusive and sustainable urbanization and capacity for participatory, integrated, and sustainable human settlement planning and management in all countries (11.3).
- *Natural and cultural heritage*. A special emphasis is put on the protection of natural and cultural heritage (11.4).
- *Poor and vulnerable citizens*. It aims to decrease the overall number of deaths and to protect the poor and vulnerable (11.5).
- *Environmental impact*. It is directed toward reducing the environmental impact of cities, with a special emphasis on air quality and waste reduction in cities and communities (11.6).
- *Disadvantaged groups*. The target aims at providing access to safe, inclusive, and accessible green public spaces for disadvantaged groups (children, women, the elderly, and people with disabilities) (11.7).

These targets of SDG 11 are being implemented in an uncertain and complex environment, where many unsolved questions arise: Is there enough water? (e.g., SDG 11.6); Is there sufficient energy? (SDG 11.6); Will there be safe and accessible public transport for all citizens? (SDG 11.2); What are the proper policy strategies that meet the challenges of societies? (SDG 11.3). Traditional planning methods may prove unsatisfactory in dealing with the changes that world societies face nowadays. By applying foresight exercises to all proposed targets, the relevance of applied policies will increase, as well as the number of (unexpected) opportunities. This normative approach allows decision-makers to prepare better and to shape the future in desirable ways. Therefore, SDGs policies will become more resilient and effective in a broader range of circumstances. Moreover, foresight exercises have broad areas of application, starting from education and technology to business, transport, and environment. That is why many national agencies, government planners, and corporate strategists use vision forecasting to produce important policy adjustments or enhancements that contribute to overall well-being.

For 2030 Agenda to meet its strategic goals, a strong and collaborative vision of the future is a suitable solution. It will help communities, businesses, and civil society as a whole to identify a suitable path and to use effectively the available resources.

As we can see from these four different lists of themes, the common focus is on city governance, living in the city, city economy, access to open data and transparency, and environmental protection. From this common ground, we can draw the conclusion that a smart city shares its typical themes with any other city but adds a particular feature, the special focus on data sharing and transparency in order to create information flows making it smart.

3.2 Imagining the future: Mindsets, activities, and techniques used in city foresight

City foresight is not a standardized activity, allowing the use of a wide array of activities and techniques to imagine the future. This is permitted by the different mindsets behind the wide choice of activities and techniques available for looking at the far future, each of them with a potential to reveal a different perspective on the future (Rogers & Hunt, 2019). Based on the fact that they start from totally different aspects of the future, we considered the following techniques for looking into the future:

- Trend analysis—It is an essential part of a design strategy and a well-known analytical tool as it helps identify universal features, useful for shorter-term planning by using different assumptions as a starting point. Through quantitative and/or qualitative methods of analysis, trends provide reliable and extensive patterns of change based on the extrapolation of historical data tendencies into the future. The method of analysis is from the periphery to center.
- Horizon scanning—It represents a systemic examination of information for spotting likely threats, risks, emerging issues, and opportunities within the time horizon set by the foresight exercise (Curry, Hodgson, Kelnar, & Wilson, 2005; Rogers & Hunt, 2019). This kind of examination is commonly

applied in foresight exercises, revealing how emerging trends might affect current policy and practice.

- Side-swipes or black swans—This kind of thinking is looking for the so-called "wildcards," and characterizes an unpredictable event that is beyond what is normally expected in a given situation. This type of event has a severe impact and is extremely rare.
- Scenario analysis—It is an exploration at the edge of societal "plausibility space," keeping the most recognizable aspects in the way societies operate, and describing situations in which a dramatic change of one aspect of societal behavior typically dominates (Rogers & Hunt, 2019).
- "Drivers matrix"—This foresight approach requires picking up two drivers of change connected to the investigated topic. Out of change-makers—one produces a high impact, while the other is of high uncertainty. By plotting these drivers of change orthogonally, we create four quadrants which allow us to explore the likely future based on the positions on the axes.
- "Aspirational scenarios"—It is a special approach to scenario analysis as it uses an emotional perspective in defining a scenario built on conventional expectation (i.e., trend analysis), growing desperation (based on a list of major city challenges), or high aspiration (based on visionary strategies leading to successful outcomes) (Rogers & Hunt, 2019).

If we look at the nature of activities in a city foresight, we can differentiate them in terms of the type of knowledge processing about the future and the result of that processing. In the following paragraphs, we highlight how the list of activities in a city foresight, presented by GOS (2017), meets precisely these criteria of classification:

Visioning activities. These explorative or normative activities comprise creative and divergent thinking in articulating aspirations and convergent thinking for developing visions and identifying value-based goals. The goal of these activities is to supply directions throughout the foresight expedition and to finally help in defining practical goals and outcomes for long-term development.

Analyzing activities. These activities are intense in their use and integration of intelligence derived from factual data about the past, the present, and the future. Risks, constraints, and opportunities can be highlighted. These require analytical thinking in identifying parameters that can affect the future supply and demand for services, physical constraints, demographic forecasts, infrastructure capacity, or financial resources. The goal of these activities is to recognize the signs of future changes in the currently available information.

Designing activities. These are activities involving a high level of creativity for developing new possibilities and options for the future. Reuse of past experience is limited, while creative imagination and development of alternative trajectories for future change are required. The goal is to create mental constructs about the future, draw the sketch of the grand design of the potential futures (set of scenarios) that can match the aspirations of stakeholders and, at the same time, respect the analytical constraints and opportunities anticipated as drivers of change.

Testing activities. The raw sketches of the future are usually generated with an open mind focused on divergent thinking. The results might be interesting, but rational thinking must be used to assess their realism. This is the reason way in foresight there is always a need for testing the newly conceived solutions or perspectives against the potential risks or constraints in order to see their robustness when confronted with a diverse range of potential uncontrollable contexts.

Assembling activities. These activities are to be found all over the foresight exercises when associative thinking is required. The usual high level of complexity and uncertainty in the information and knowledge processed during foresight exercises is a fertile ground for attempting to associate new, disparate concepts to develop new perspectives and solutions from previously isolated silos. These activities take place for creating new delivery models, technological innovation, financing solutions, institutional reforms or regulatory development, new skills mapping, etc. The general purpose of these activities is to assemble complex parts, resulted from different forms of foresight thinking, in order to create coherent and plausible roadmaps (GOS, 2017).

3.3 Main benefits of foresight for smart cities

There are many reasons why foresight is important for smart cities, but the most significant benefits relate to valuing local strengths, fostering the relationships between the stakeholders, strategy refinement, business environment confidence, emergence of new networking initiatives and strategic partnerships, solving chronic problems, and tackling new challenges and citizen engagement.

One of the most important potential benefits of foresight for creating a smart city might be a refreshing of the city brand by assessing the local strengths. The stakeholders involved in the foresight add their fresh perspectives to unique local assets by creating emerging trends for new potential ways to create value for the city, combining them into distinct features that can define a new city brand or identity in the future. Ideas challenging the status quo can be particularly powerful for cities struggling to find a new brand and identity. The smart city vision might be one of the new features of brand identity.

Another benefit of a city foresight is fostering the relationships among the stakeholders brought together at the same table, to deal with the same topics. It enhances the future resilience of the city by building organic relations between the main city actors, useful to cope with change and disruption, if needed. In a smart city, the level of expected interconnectivity between stakeholders is very intense, therefore, the use of a foresight exercise might be the best thing you can do to start this process of intensifying the interconnectivity.

Strategy refinement is a major benefit of a city foresight, typical to all foresight projects. The collaborative approach of foresight brings a deeper sense of alignment of local aspirations, institutional missions, and stakeholders' goals, clarifying the risks and the steps of potential roadmaps toward the future and helping the design of more robust and complex strategies. "Increased clarity and alignment of place-specific aspirations, local assets, opportunities, and enhanced awareness of risks yield more robust strategic 'roadmaps' towards the future" (Gov., UK, 2017, p. 8). Local governance institutions have the opportunity to discover valuable insights into the intricate policy relationships, thus enhancing their integrative and systemic thinking. Also, early identification of risks might result from divergent future-oriented thinking used in the environmental analysis.

Business environment confidence is another benefit of foresight for a smart city. The public confidence in city evolution under a leadership being concerned about long term development brings a supplementary sense of confidence for the investors, enhancing the city attractiveness to businesses and skilled workers. Economic prosperity is not only achieved by public policies, resulted from foresight, but it is enhanced by this hidden effect. In the medium-term, a smart city creating a positive perception of the quality of its "roadmap" toward the future is likely to be appealing as a future place of living and working for highly skilled workers.

Other benefits of city foresight are the emergence of new networking initiatives and strategic partnerships. The collaborative exercises, leading to the identification of future business and development opportunities across wider geographies are other opportunities to develop new connections, start collaborative networks, and build coalitions. The trans-functional nature of smart city applications asks for approaches of networking type and involvement in strategic partnerships for ensuring functionality and real benefits for the city. Insights and data gained from city foresight may also be used as arguments for strengthening advocacy at the central government for attracting central funding for specific smart city projects.

Solving chronic problems and tackling new challenges is also a recognized benefit of city foresight. In the case of a smart city project, the space created for expressing the aspiration of relevant stakeholders and the wide spectrum of possibilities opened by disconnecting the thoughts from current resource limitations and behavioral restrains are facilitators for bringing into discussion chronic problems and for being bold enough to tackle significantly difficult challenges by complex solutions using new technologies and new coalitions. The foresight projects are recognized as excellent learning environments by all actors and are seen as expressions of the city transforming itself into a learning organization.

The gained civic engagement could be one of the most important gains of a foresight project used for fostering the evolution toward a smart city. It is achieved by means of collaborative activities involving the participation of stakeholders and by public dissemination of the results of the foresight. High creative engagement of the general public provides opportunities to display the city leadership qualities and enhance civic pride. This participation can provide a greater buy-in for future decisions coherent with the vision, scenarios, or roadmaps disseminated to the general public (Gov. UK, 2017). Also, this may be one of the behavioral building blocks in creating a smart city citizen culture, where the citizen is conscientious about its role in the complex mechanisms of a smart city.

4. Conclusions

Foresight exercises provide solutions to support both governments and citizens in a lasting and adaptable way, and in line with the goals of the 2030 Agenda.

The themes of the foresight exercises have been influenced worldwide by the 11th goal of the 2030 Agenda for Sustainable Development, focused on transforming cities and human settlements into more inclusive, safe, resilient, and sustainable societies. The statements of SDG 11 are used as a reference for the creation of smart city visions in foresight exercises. The foresight project for a smart city should be positioned from the beginning as being a community-owned project providing full access to relevant data to all significant stakeholders and generating results for the benefit of the entire city community. Accountability for the success of the project is usually assumed by the policymakers involved in the process as they are the ones mainly responsible for turning the foreseen future into reality.

A foresight project usually uses a horizon of 15–25 years into the future. This time span allows stakeholders to free their thoughts from the main restraints of the present, making them able to identify potential opportunities and likely future developments. The engagement of stakeholders is paramount for city foresight. To understand the technological developments and the subsequent social consequences of the city's evolution, it is necessary to assemble a multidisciplinary smart city team having members of various backgrounds: technological, social, political, and educational. This team will build a vision from a multidisciplinary perspective, pushing acceptance further for the implementation of the new policies.

The smart city foresight project should use external knowledge sources, especially from other smart cities where relevant best practices communities can offer valuable insights. The foresight working teams can benefit from global knowledge transfer and understanding of working methods shared by these communities or networks.

The smart city is based on interoperability of data systems and this interoperability has its seeds in the foresight exercises. The smart city team should ensure that all actors are able to provide complementary smart city information, first, during the foresight exercise, and later, continuously, in order to maintain the functioning of the cybernetic loops of the city organism. Any monopoly over any public data should be replaced by transparency and sharing. Implementation of data-driven initiatives is one of the major sources of the functionality of a smart city.

Stakeholders should be involved in identifying the main axes of the city development and engaged in the foresight process. They are the backbone of the new emerging scenario. They will determine the methodological choices and will be used as criteria for assessing the practical value of the outcomes of different solutions. Not least, they will represent the main Public Relation topics when disseminating the results of the foresight. In most cases, the axes converge around the interconnection between the use of IT&C, city governance, city economy and environmental sustainability, living and working, city transportation, and infrastructure.

The city complexity conceptualization, achieved during the foresight exercise, is considered a significant determinant of the foresight success. The smart city is usually characterized by a high level of functional and contextual complexity. The determinants of coherent foresight results, capable of influencing policy-making, are the deep understanding of smart city complex elements and the extent to which the details of these elements are taken into consideration during the foresight.

The foresight project should use proven methods of foresight adapted for the particular context of a city. Foresights experts should be involved in creating a mix of methods that take into account the need to combine visioning, analysis, design, testing, and assembling of activities in order to achieve realistic scenarios and creative solutions.

The partial and the final results of the foresight project must be disseminated as much and as often as possible to as many stakeholders as possible. Besides the experts, engineers, academics, business owners, or policy-makers, many of the new initiatives will involve almost all citizens and their approval and behavioral change will be the key to success. The future, being a consequence for everybody in a city, depends on active responsibility assumed by the entire community.

References

Anthopoulos, L. (2015). Understanding the smart city domain: A literature review. In M. Rodríguez-Bolívar (Ed.), *Public administration and information technology. Transforming city governments for successful smart cities*. Switzerland: Springer International Publishing.

Cetindamar, D., Lammers, T., & Sick, N. (2020). Digital technologies, competitiveness & policies: An integrative city-based policy roadmap for entrepreneurial ecosystems. In *Managing innovation in a global and digital world* (pp. 49–62). Wiesbaden: Springer.

Coates, J. F. (1985). Foresight in federal government policy making. *Futures Research Quarterly*, *1*(2), 29–53.

Curry, A., Hodgson, A., Kelnar, R., & Wilson, A. (2005). *Intelligent infrastructure futures: The scenarios—Towards 2050*. Report commissioned for the Foresight Programme of The Office of Science and Technology, London, UK.

Dixon, T., Montgomery, J., Horton-Baker, N., & Farrelly, L. (2018). Using urban foresight techniques in city visioning: Lessons from the Reading 2050 vision. *Local Economy*, *33*(8), 777–799.

Duinker, P. N., Ordóñez, C., Steenberg, J. W., Miller, K. H., Toni, S. A., & Nitoslawski, S. A. (2015). Trees in Canadian cities: Indispensable life form for urban sustainability. *Sustainability*, *7*(6), 7379–7396.

Duranton, G., & Puga, D. (2001). Nursery cities: Urban diversity, process innovation, and the life cycle of products. *American Economic Review*, *91*(5), 1454–1477.

Farole, T., Rodríguez-Pose, A., & Storper, M. (2009). *Cohesion policy in the European Union: Growth, geography*. Institutions Report. Working paper of London School of Economics.

Fernández Güell, J. M., & López, J. G. (2016). Cities futures. A critical assessment of how future studies are applied to cities. *Foresight*, *18*(5), 454–468.

FOREN (Foresight for Regional Development Network). (2001). A practical guide to regional foresight. In J. Gavigan, & F. Scapolo (Eds.),. Brussels, BE: European Commission.

GOS (Government Office for Science). (2017). *Future of cities: Foresight for cities*. Retrieved from www.gov.uk/government/publications/future-of-cities-foresight-for-cities.

Gov., UK. (2017). *The futures toolkit: Tools for futures thinking and foresight across UK government*. London, UK: Government Office for Science.

Hoq, M. N., Alam, R., & Amin, A. (2019). Prediction of possible asthma attack from air pollutants: Towards a high density air pollution map for smart cities to improve living. In *2019 International Conference on Electrical, Computer and Communication Engineering (ECCE), Cox'sBazar, Bangladesh* (pp. 1–5). IEEE. https://doi.org/10.1109/ECACE.2019.8679335.

John, B., Keeler, L. W., Wiek, A., & Lang, D. J. (2015). How much sustainability substance is in urban visions? An analysis of visioning projects in urban planning. *Cities*, *48*, 86–98.

Komninos, N., Bratsas, C., Kakderi, C., & Tsarchopoulos, P. (2019). Smart city ontologies: Improving the effectiveness of smart city applications. *Journal of Smart Cities*, *1*(1), 31–46.

Marinova, S., Ul-Haq, R., Portaleoni, C. G., & Marinov, M. (2013). *Corporate foresight and strategic decisions: Lessons from a European bank*. Basingstoke: Palgrave Macmillan.

Martin, B. R. (1995). Foresight in science and technology. *Technology Analysis & Strategic Management*, *7*(2), 139–168.

Martin, B. R. (2010). The origins of the concept of 'foresight' in science and technology: An insider's perspective. *Technological Forecasting and Social Change*, *77*(9), 1438–1447.

Matos, F., Vairinhos, V. M., Dameri, R. P., & Durst, S. (2017). Increasing smart city competitiveness and sustainability through managing structural capital. *Journal of Intellectual Capital*, *18*(3), 693–707.

McPhearson, T., Iwaniec, D., & Bai, X. (2016). Positive visions for guiding urban transformations towards sustainable futures. *Current Opinion in Environmental Sustainability*, *22*, 33–40.

Nam, T., & Pardo, T. A. (2011). Conceptualizing smart city with dimensions of technology, people, and institutions. In *Proceedings of the 12th Annual International Digital Government Research Conference: Digital Government Innovation in Challenging Times* (pp. 282–291). New York, NY: ACM International Conference Proceeding Series, Association for Computing Machinery. https://doi.org/10.1145/2037556.2037602.

Nenciu, G., Van Der Duin, P., Janssen, M., Snijders, D., & Hulstijn, J. (2017). Future scenarios of data-driven smart cities. In *Presented at the 18th Annual International Conference on Digital Government Research*. Available at https://stt.nl/stt/wp-content/uploads/2016/08/Future-Scenarios-of-Data-Driven-Smart-Cities.pdf. (Accessed 23 September 2018).

Norton, B. A., Coutts, A. M., Livesley, S. J., Harris, R. J., Hunter, A. M., & Williams, N. S. (2015). Planning for cooler cities: A framework to prioritise green infrastructure to mitigate high temperatures in urban landscapes. *Landscape and Urban Planning*, *134*, 127–138.

Pérez-del Hoyo, R., Andújar-Montoya, M. D., Mora, H., & Gilart-Iglesias, V. (2019). Unexpected consequences in the operation of urban environments. *Kybernetes*, *48*(2), 253–264. https://doi.org/10.1108/K-02-2018-0096.

Phillips, F., Chang, J., & Su, Y. S. (2019). When do efficiency and flexibility determine a firm's performance? A simulation study. *Journal of Innovation & Knowledge*, *4*(2), 88–96.

Ravetz, J., & Miles, I. (2016). Foresight in cities: On the possibility of a "strategic urban intelligence". *Foresight, 18*(5), 469–490. https://doi.org/10.1108/FS-06-2015-0037.

Reger, G. (2001). Technology foresight in companies: From an indicator to a network and process perspective. *Technology Analysis & Strategic Management, 13*(4), 533–553.

Rogers, C. D. F. (2017). The value of foresight and scenarios in engineering liveable future cities. In M. Eames, T. Dixon, M. Hunt, & S. Lannon (Eds.), *Retrofitting cities for tomorrow's world* (pp. 139–152). Chichester, UK: Wiley Blackwell.

Rogers, C. D., & Hunt, D. V. (2019). Realising visions for future cities: An aspirational futures methodology. *Proceedings of the Institution of Civil Engineers-Urban Design and Planning, 172*(4), 125–140.

Schaffers, H., Komninos, N., Pallot, M., Trousse, B., Nilsson, M., & Oliveira, A. (2011). Smart cities and the future internet: Towards cooperation frameworks for open innovation. In *The future internet assembly* (pp. 431–446). Berlin, Heidelberg: Springer.

UN (United Nations). (2015). *Transforming our world: The 2030 agenda for sustainable development.* Retrieved from https://sustainabledevelopment.un.org/post2015/transformingourworld/publication.

United Nation Economics and Social Council (UNESC). (2015). *The UNECE–ITU smart sustainable cities indicators.* Retrieved from http://www.unece.org/fileadmin/DAM/hlm/documents/2015/ECE_HBP_2015_4.en.pdf.

Urban Foresight. (2016). *Smart cities Scotland blueprint—Report for Scottish Cities Alliance.* Retrieved from https://www.scottishcities.org.uk/site/assets/files/1103/smart_cities_scotland_blueprint_web-4.pdf.

Van der Meulen, B., De Wilt, J., & Rutten, H. (2003). Developing futures for agriculture in the Netherlands: A systematic exploration of the strategic value of foresight. *Journal of Forecasting, 22*(2–3), 219–233.

Visvizi, A., & Lytras, M. (Eds.). (2019). *Smart cities: Issues and challenges: Mapping political, social and economic risks and threats.* Amsterdam: Elsevier.

Voda, A. I., & Radu, L. D. (2018). Artificial intelligence and the future of smart cities. *Brain (Broad Research in Artificial Intelligence and Neuroscience), 9*(2), 110–127.

Wippel, N. (2014). *Multinational strategic foresight of environmental trends in the automobile industry using internal resources. Vol. 9.* Münster, Germany: LIT Verlag.

Zeng, M. A., Koller, H., & Jahn, R. (2019). Open radar groups: The integration of online communities into open foresight processes. *Technological Forecasting and Social Change, 138*, 204–217.

Chapter 16

Transforming the city from within: SDGs and their transformative dynamics in the smart city (conclusions)

Anna Visvizi[a,b] and Raquel Pérez del Hoyo[c]
[a]*SGH Warsaw School of Economics, Warsaw, Poland,* [b]*Deree College, The American College of Greece, Athens, Greece,* [c]*University of Alicante, Alicante, Spain*

1. Introduction

By now, smart city is a concept and policy-making imperative that is well established in the academic debate and in the policy-making agenda. Several approaches to the study of smart cities exist and offer a frequently very critical insight into issues and developments that shape the debate on smart cities and the urban space itself (Augusto, 2020; Coletta, Evans, Heaphy, & Kitchin, 2019; Visvizi & Lytras, 2019). Disciplinary boundaries play a role in the debate on the smart city, and so do personal views and standpoints. This volume seeks to contribute to the debate by highlighting the ways in which the smart city and the Sustainable Development Goals (SDGs) Agenda connect and therefore benefit the smart city and its inhabitants. Purposefully multidisciplinary, this volume offers a detailed insight into a selection of issues that are important in the debate on the smart city and may be useful not only for researchers but also for people directly involved in the decision-making process at the smart city level. Smart city services and applications made their way into the fabric of the modern city gradually, in some cases, however, diametrically transforming the city inhabitant's experience of the city. Frequently, but not always though, the transformation of a city into a smart city is a direct function of the development of e-government solutions. The cases discussed in this volume highlight it.

2. The key points and questions raised in this volume

The chapters included in this volume apply the lens of the SDGs, especially SDG 11, to address several general and specific issues that shape the debate

Smart Cities and the UN SDGs. https://doi.org/10.1016/B978-0-323-85151-0.00016-6

on smart cities today. Case studies have been used throughout the book to ensure the connection between the conceptual and the empirical insight employed in the analysis. As a result, this volume may serve as a primer for all those who wish to understand the connection between the imperatives inherent in the SDGs and the promise entailed in the concept of the smart city. Advances in ICT, including big data, data analytics, data mining, sensors, virtual reality, augmented reality, 5G technologies, etc., redefine the landscape of city/urban space. They have a profound impact on how we live, work, spend our free time, and travel. They influence the way we receive our medical treatment, the degree of safety (not only online) we get, and the way and the frequency we communicate, etc. Applied in the city space, these technologies inadvertently transform cities into smart cities. The chapters comprising this volume make a case for a multifaceted, ICT-conscious, comprehensive, and integrated strategy aimed at advancing a smart city agenda compliant with the imperative of inclusion, safety, resilience, and sustainability. This implies an emphasis on participation and representation, cocreation, and smart governance. Technology in this view retains its role as a means necessary to achieve that.

The chapters included in this volume have been structured in four parts, addressing the conceptual issues specific to the SDGs and the smart city; ways and means of involving citizens in processes related to the functioning and development of the smart city; the interconnected notions of safety, security, and well-being in the smart city; and finally delving into some of the sociological aspects underlying the functioning of the smart city, including the lifestyle, and knowledge sharing. What brings these chapters together is their emphasis on SDGs, especially SDG 11. The contributing authors shed light on diverse ways in which the smart city and the SDGs can mutually reinforce each other.

This edited volume addresses several big questions that shape the debates, respectively, on smart cities and the SDGs. The key issues and questions that the contributing authors address include: Which tools, which means, which policies to apply to support cities in the process of attaining sustainability? What are the key obstacles to the implementation of a sustainability-oriented growth and development agenda in the smart city? What are the key structural problems that cities are exposed to and how can the imperative of sustainability facilitate the process of bypassing them?

The chapters included in this volume address also such questions as in which ways the making of smart cities can improve people's well-being? To what extent can insights from sociology, political science, and economics boost the process of attaining SDG 11? What is the role and place of urban planning in the debate on smart cities? What is the role of national and local stakeholders in the process of developing a smart and sustainable city? How should ICT be conceived in the process of making cities inclusive, safe, resilient, and sustainable? To what extent and how can innovation and technology boost strategies conducive to citizens' well-being?

This list of questions is by no means exhaustive. Several other questions have been addressed in the volume, including, for instance: open data portal in the smart city context, cybersecurity issues and ways of addressing them, cocreation in the smart city; e-democracy and digital governance in the smart city. In addition, this edited volume explores questions related to sustainability assessment, i.e., the foundations and principles behind measuring sustainability, quality of smart infrastructure. For instance, what set of criteria should be measured to assess the level of housing intelligence? What alternatives can be chosen to define a foresight exercise prior to practical initiatives to transform city systems and public governance policies to become a smart city? In brief, this volume offers a reflection-driven insight into the use of sophisticated ICT in (smart) city/urban space in a period when the imperative of sustainability is not a matter of a choice anymore.

3. The way forward: Let's be pragmatic

The content of this volume suggests that several topics still require greater attention on the part of the research community. These issues and topics include: the need to raise awareness in the society and among decision-makers that technology-supported smart cities are key to the sustainable development of cities and the achievement of SDGs. It is equally important to promote research that engages conceptual and empirical insights, ideally also in a comparative manner, to understand the mechanisms enabling participation, cocreation, and governance in the smart city. The notion of knowledge generation and knowledge sharing, also in terms of innovation development and promotion, constitutes another salient area of research that deserves continued attention. What follows is that the smart cities research could benefit from insights based on information management. That is, the key question here is how to effectively collect, manage, and apply information collected from the smart city inhabitants and stakeholders, i.e., including also businesses operating in and using the smart city infrastructure (Lytras & Visvizi, 2021).

Today's cities, regardless of whether they have the capacity and resources to apply smart city solutions or not, face a variety of mutually reinforcing challenges, including resource management, irregular migration, energy, water scarcity, and many more (cf. Chui, Lytras, & Visvizi, 2018; Visvizi, Pachocka, & Lytras, 2019). The UN's focus on cities and the inclusion of SDG 11 in the SDG Agenda prove that the centrality of cities and urban systems is recognized by the decision-makers globally. However, the case of SDG 11, discussed throughout the chapters included in this volume, serves also as an excellent example that proves the value of multilateralism and of global governance. Multilateralism is at its best in this case because the 2030 Sustainable Development Agenda is the outcome of deliberation and consensus reached in the forum of the United Nations (cf. Surdej, 2020). Global governance comes in this picture too, however, because the mechanisms by means of which the SDGs are implemented are more nuanced than international collaboration. The successful implementation of SDG 11 necessitates market-driven tools and mechanisms that allow nonstate actors to become agents of change too (Visvizi, 2015). The case of the smart city is a good example here suggesting that the city authorities, for lack of powers and resources, may not necessarily be the only actors involved in promoting inclusion, safety, resilience, and sustainability (cf. Clarinval, Simonofski, Vanderose, & Dumas, 2020). As the authors contributing to this volume suggest, other stakeholders, such as the business sector, vendors of services, citizens themselves, and other mechanisms, such as cocreation, may well complement actions taken by the traditional actors, i.e., the state and the local authorities (cf. Troisi, D'Arco, Loia, & Maione, 2018).

The problem inherent in the smart city debate is that considerable differentiation in the spread and application of ICT-enhanced solutions in the city space exists. In other words, several places in the world exist where the lack of basic infrastructure stands in the way to the adoption of smart city services. In some places, smart city services may be adopted, but people, due to lack of skills or lack of a device, are denied access to it. Examples could be multiplied. The point is, however, that research on smart cities, regardless of the disciplinary origin and/or ideological orientation, ought to stress that a gap exists between what is technically possible and what is feasible on the ground. This gap and its countless social, geographical, economic, and political facets ought to be highlighted over and over again, if we—as the research community, scholars, and educators—want to claim to be serious about SDGs and SDG 11.

References

Augusto, J. C. (Ed.). (2020). *Handbook of smart cities*. Cham: Springer. https://doi.org/10.1007/978-3-030-15145-4.

Chui, K. T., Lytras, M. D., & Visvizi, A. (2018). Energy sustainability in smart cities: Artificial intelligence, smart monitoring, and optimization of energy consumption. *Energies, 2018*(11), 2869. https://doi.org/10.3390/en11112869.

Clarinval, A., Simonofski, A., Vanderose, B., & Dumas, B. (2020). Public displays and citizen participation: A systematic literature review and research agenda. *Transforming Government: People, Process and Policy*. https://doi.org/10.1108/TG-12-2019-0127.

Coletta, C., Evans, L., Heaphy, L., & Kitchin, R. (Eds.). (2019). *Creating smart cities*. London and New York: Routledge.

Lytras, M. D., & Visvizi, A. (2021). Information management as a dual-purpose process in the smart city: Collecting, managing and utilizing information. *International Journal of Information Management, 56*. https://doi.org/10.1016/j.ijinfomgt.2020.102224.

Surdej, A. (2020). Multilateralism and international governmental organizations: Principles and instruments. *Transforming Government: People, Process and Policy, 14*(3), 337–350. https://doi.org/10.1108/TG-11-2019-0107.

Troisi, O., D'Arco, M., Loia, F., & Maione, G. (2018). Big data management: The case of Mulino Bianco's engagement platform for value co-creation. *International Journal of Engineering Business Management*, (January). https://doi.org/10.1177/1847979018767776.

Visvizi, A. (2015). Safety, risk, governance and the Eurozone crisis: Rethinking the conceptual merits of 'global safety governance'. In P. Kłosińska-Dąbrowska (Ed.), *Essays on global safety governance: Challenges and solutions* (pp. 21–39). Warsaw: Centre for Europe, University of Warsaw, ASPRA-JR, ISBN:83-89547-24-4.

Visvizi, A., & Lytras, M. D. (Eds.). (2019). *Smart cities: Issues and challenges: Mapping political, social and economic risks and threats* Elsevier, ISBN:9780128166390. https://www.elsevier.com/books/smart-cities-issues-and-challenges/lytras/978-0-12-816639-0.

Visvizi, A., Pachocka, M., & Lytras, M. D. (2019). Managing international migration: Rethinking transaction costs, red tape, and their impact. *International Migration*, *57*(5), 271–279. https://doi.org/10.1111/imig.12648.

Index

Note: Page numbers followed by *f* indicate figures and *t* indicate tables.

Q

R

S

FSC
www.fsc.org
MIX
Papier | Fördert
gute Waldnutzung
FSC® C083411

Kolibri 360 GmbH
Ferdinand-Jühlke-Straße 7
99095 Erfurt, Deutschland
produktsicherheit@kolibri360.de